URBAN WATER SECURITY

Challenges in Water Management Series

Editor:

Justin Taberham

Independent Consultant and Environmental Advisor, London, UK

Other titles in the series:

Water Resources: A New Management Architecture

Michael Norton, Sandra Ryan and Alexander Lane

2017

ISBN: 978-1-118-79390-9

URBAN WATER SECURITY

ROBERT C. BREARS

WILEY

This edition first published 2017 © 2017 by John Wiley & Sons, Ltd

Registered Office
John Wiley & Sons, Ltd, The Atrium, Southern Gate, Chichester, West Sussex, PO19 8SQ, UK

Editorial Offices
9600 Garsington Road, Oxford, OX4 2DQ, UK
The Atrium, Southern Gate, Chichester, West Sussex, PO19 8SQ, UK
111 River Street, Hoboken, NJ 07030-5774, USA

For details of our global editorial offices, for customer services and for information about how to apply for permission to reuse the copyright material in this book please see our website at www.wiley.com/wiley-blackwell.

Library of Congress Cataloging-in-Publication data applied for

ISBN: 9781119131724

A catalogue record for this book is available from the British Library.

Wiley also publishes its books in a variety of electronic formats. Some content that appears in print may not be available in electronic books.

Cover image: © Peter Zelei images/Gettyimages

Set in 10/12pt Melior by SPi Global, Pondicherry, India

Printed in Singapore by C.O.S. Printers Pte Ltd

10 9 8 7 6 5 4 3 2 1

Contents

Series Editor Foreword – Challenges in Water Management

The World Bank in 2014 noted:

> Water is one of the most basic human needs. With impacts on agriculture, education, energy, health, gender equity, and livelihood, water management underlies the most basic development challenges. Water is under unprecedented pressures as growing populations and economies demand more of it. Practically every development challenge of the 21st century – food security, managing rapid urbanization, energy security, environmental protection, adapting to climate change – requires urgent attention to water resources management.
>
> Yet already, groundwater is being depleted faster than it is being replenished and worsening water quality degrades the environment and adds to costs. The pressures on water resources are expected to worsen because of climate change. There is ample evidence that climate change will increase hydrologic variability, resulting in extreme weather events such as droughts, floods, and major storms. It will continue to have a profound impact on economies, health, lives, and livelihoods. The poorest people will suffer the most.

It is clear that there are numerous challenges in water management in the twenty-first century. In the twentieth century, most elements of water management had their own distinct set of organisations, skill sets, preferred approaches and professionals. The overlying issue of industrial pollution of water resources was managed from a 'point source' perspective.

However, it has become accepted that water management has to be seen from a holistic viewpoint and managed in an integrated manner. Our current key challenges include the following:

- The impact of climate change on water management, its many facets and challenges – extreme weather, developing resilience, storm water management, future development and risks to infrastructure
- Implementing river basin/watershed/catchment management in a way that is effective and deliverable
- Water management and food and energy security
- The policy, legislation and regulatory framework that is required to rise to these challenges
- Social aspects of water management – equitable use and allocation of water resources, the potential for 'water wars', stakeholder engagement, valuing water and the ecosystems that depend upon it

This series highlights cutting-edge material in the global water management sector from a practitioner as well as an academic viewpoint. The issues covered in the series are of critical interest to advanced-level undergraduates and masters students as well as industry, investors and the media.

Justin Taberham, CEnv
Series Editor
www.justintaberham.com

Acknowledgements

I wish to say a big thank you to all the people who took time out of their busy schedules to sit down for an interview as well as provide any supplementary material. Without your help this book would not have been possible. Specifically I wish to thank Jan Peter van der Hoek (Waternet); Jens Feddern and Joachim Jeske (Berliner Wasserbetriebe); Allan Broløs and Charlotte Storm (HOFOR); Marc Waage, Greg Fisher and Melissa Elliot (Denver Water); Christian Guenner (Hamburg Wasser); David Grantham, Karen Simpson, Paul Rutter and Rosie Rand (Thames Water); Wai Cheng Wong and Gayathri Kalyanaraman (PUB); Lisa Botticella (Toronto Water) and Jennifer Bailey (Waterworks Utility). Finally, I wish to thank mum who has a great interest in the environment and water and has supported me in this journey of writing the book.

Introduction

In the twenty-first century, the world will see an unprecedented migration of people moving from rural to urban areas: In 2012, human civilisation reached a milestone with 50 percent of the world's population living in urban settings. This is projected to reach 70 percent by 2050. With global demand for water projected to outstrip supply by 40 percent in 2030, cities will likely face water insecurity as a result of climate change and the various impacts of urbanisation.

Traditionally, urban water managers facing increased demand alongside varying levels of supplies have relied on large-scale, supply-side infrastructural projects, such as dams and reservoirs, to meet increased demands for water; however, these projects are environmentally, economically and politically costly. Environmental costs include disruptions of waterways that support aquatic ecosystems, while economic costs stem primarily from a reliance on more distant water supplies often of inferior quality. This not only increases the costs of transportation but also the cost of treatment. Furthermore, with the vast majority of water resources being transboundary, supply-side projects can create political tensions due to water crossing intra- and interstate administrative and political boundaries. As such, cities need to transition from supply-side to demand-side management to achieve urban water security.

Integrated urban water management (IUWM) recognises actions that achieve urban water security extend beyond improving water quality and managing quantity. In particular, IUWM integrates the elements of the urban water cycle (water supply, sanitation, stormwater management and waste management) into both the city's urban development process and the management of the river basin in which the city is located for the purpose of maximising water's many environmental, economic and social benefits equitably. IUWM activities to maximise these benefits include: improving water supply and consumption efficiency; ensuring adequate drinking water quality and wastewater treatment; improving economic efficiency of services to sustain operations and investments for water, wastewater

Urban Water Security, First Edition. Robert C. Brears.
© 2017 John Wiley & Sons, Ltd. Published 2017 by John Wiley & Sons, Ltd.

and stormwater management; utilising alternative water sources; engaging communities in the decision-making process of water resources management; establishing and promoting water conservation programmes; and supporting capacity development of personnel and institutions that engage in IUWM.

In IUWM, demand management is the process by which improved provisions of existing water supplies are developed. In particular, demand management promotes water conservation during times of both normal and atypical conditions through changes in practices, culture and people's attitudes towards water resources. Demand management involves communicating ideas, norms and innovative methods for water conservation across individuals and society; the purpose of demand management is to positively adapt society to reduce water consumption patterns and achieve urban water security. Demand management instruments can be divided into regulatory and technological instruments or communication and information instruments. Regulatory and technological instruments include the pricing of water, waste and stormwater to encourage water conservation as well as ensuring the efficient distribution of water. Communication and information instruments include education of young people, public awareness campaigns to encourage water conservation as well as encouraging the installation of water-efficient technologies, such as tap inserts, to reduce water consumption. The book is case study led and provides new research on the human dimensions of IUWM. In particular, it contains nine in-depth case studies of leading developed cities of differing climates, incomes and lifestyles from around the world that have used demand management tools to modify the attitudes and behaviour of water users in an attempt to achieve urban water security. Data for each case study is collected from interviews conducted with each city's respective water utility along with primary documents. The nine cities are Amsterdam, Berlin, Copenhagen, Denver, Hamburg, London, Singapore, Toronto and Vancouver. Each city scores highly on the Siemens Green City Index for water management. The Green City Index is a research project conducted by the Economist Intelligence Unit (EIU) and sponsored by Siemens. Each city is selected as a case study for the following reasons. Amsterdam is a city attracting sustainability-related companies and investments and so is attempting to manage its resources wisely while Berlin has a history of managing its water in a closed system. Copenhagen uses a variety of demand management tools to promote water conservation due to scarcity of good quality water: the majority of the city's groundwater is contaminated from agricultural and industrial production. Denver, since facing a drought in 2002, has been using demand management tools to reduce average per capita water consumption in order to increase the city's resilience to future droughts. Hamburg has a history of relying on imported water but faces population growth challenges. Similarly, London has implemented demand management efforts in response to demand outstripping supply due to rapid population growth, along with a changing climate. Singapore has a limited surface area to collect surface water and has no groundwater supplies; hence, the city state imports nearly all of its water from neighbouring Malaysia. To reduce the country's dependency on imported water, the city has implemented aggressive water conservation campaigns in an attempt to achieve urban water security. Toronto, despite being located by the Great Lakes, has implemented water conservation efforts in response to the city government requiring its utilities to be sustainable, both environmentally and financially. Finally, Vancouver is implementing demand management strategies to ensure the city does not have to expand its storage capacity to meet rising demand.

This book will introduce readers to the transition management framework that guides cities and their transitions towards urban water security through the use of demand management strategies. A transition in IUWM is a well-planned, coordinated transformative shift from one water system to another, over a long period of time, where a water system comprises physical and technological infrastructure, cultural/political meanings and societal users. In a water system, society is both a component of the water system and a significant agent of change in the system, both physically (change in processes of the hydrological cycle) and biologically (change in the sum of all aquatic and riparian organisms and their associated ecosystems). In IUWM, transitions to new water systems are triggered by changes in the external environment of the system, leading to it being inefficient, ineffective or inadequate in fulfilling its societal function: the main drivers of water insecurity are rapid population and economic growth, increased demand for food and energy and climate change. In transitions towards urban water security, cities set a target water consumption level to achieve (per capita litres/day, for example) with the baseline for comparison being current levels of water consumption and select a portfolio of demand management tools to promote the better use of existing water supplies before plans are made to further increase supply. Overall, transitions in IUWM involve an iterative, long-term and continuous process of influencing people's beliefs and practices to achieve urban water security.

The importance of this book is that in IUWM our understanding of the social, economic and political dimensions of demand for water lags significantly behind engineering and physical science knowledge on the supply of urban water resources. As such, little has been written on the actual processes that enable the application of IUWM; therefore, it is difficult to demonstrate or compare successes across cities in managing urban water sustainably. This is despite the fact it is human attitudes and behaviour that determines the actual amount of water that needs supplying. More specifically, the emphasis on engineering, scientific and technological solutions is no longer sufficient to deal with the numerous problems and uncertainties of increasing demand and climate change on water resources. Therefore, it is critical that human dimensions are incorporated into the managing of urban water, as the perspective of society is crucial for the success or failure of any water management strategy. Nevertheless, the concept of IUWM for addressing water scarcity is changing only slowly from an emphasis on science and technology towards solutions that incorporate cultural and behavioural change. This book presents new research on the human dimensions of IUWM. In particular, the book is case study led containing nine case studies on how leading developed cities from around the world have used demand management strategies (involving regulatory and technological and information and communication instruments) to modify the attitudes and behaviour of water users in an attempt to achieve urban water security. Each case study is written from the perspective of the water utility with input from each city's respective water utility representative.

The book's chapter synopsis is as follows:

Chapter 1 provides a 'Water 101' for readers to understand what exactly constitutes water and how the quality and quantity of water can vary naturally. The chapter will then describe the impacts of urbanisation on water quality and quantity.

Chapter 2 defines what water security is and the challenges to achieving urban water security. These challenges include rapid economic and population growth, urbanisation and rising demand for energy and food as well as climate change.

Chapter 3 defines what sustainability and sustainable development is before discussing the differing approaches to sustainability. The chapter introduces sustainable water management frameworks to achieve water security and then discusses how IUWM can achieve urban water security by balancing demand for water with supply.

Chapter 4 first discusses the purpose of demand management strategies before discussing the types of demand management strategies available to urban water managers. The chapter then discusses demand management tools available to water managers in transitions towards urban water security.

Chapter 5 provides readers with a definition of a transition before discussing types of transitions, how they occur over and the various drivers and forces of transitions. The chapter then discusses how transitions can be managed.

Chapter 6 discusses transitions in the context of managing natural resources sustainably. In particular, the chapter discusses transitions in the context of climate change and natural resource scarcity before introducing readers to transitions towards the sustainable management of water to achieve urban water security.

Chapter 7 provides readers with a case study on Amsterdam transitioning towards urban water security through demand management.

Chapter 8 provides readers with a case study on Berlin transitioning towards urban water security through demand management.

Chapter 9 provides readers with a case study on Copenhagen transitioning towards urban water security through demand management.

Chapter 10 provides readers with a case study on Denver transitioning towards urban water security through demand management.

Chapter 11 provides readers with a case study on Hamburg transitioning towards urban water security through demand management.

Chapter 12 provides readers with a case study on London transitioning towards urban water security through demand management.

Chapter 13 provides readers with a case study on Singapore transitioning towards urban water security through demand management.

Chapter 14 provides readers with a case study on Toronto transitioning towards urban water security through demand management.

Chapter 15 provides readers with a case study on Vancouver transitioning towards urban water security through demand management.

Chapter 16 provides readers with a series of best practices and lessons learnt from the selected case studies of water utilities implementing demand management strategies in an attempt to achieve urban water security. The chapter then provides readers with a range of recommendations to achieve further urban water security.

1 Water 101

Introduction

Before we can manage water sustainably to achieve water security – in the face of global challenges including rapid economic and population growth, rising demand for energy and food and climate change impacting the availability of water resources – we need to understand what is water and its natural variations in terms of quantity and quality. This chapter will first describe the physical properties of water, before discussing the Earth's hydrological cycle. The chapter will then discuss natural variations to water quantity and water quality before finally providing readers with an overview of the impacts of urbanisation on water resources.

1.1 What is water?

On Earth, 97.5 percent of all water is saltwater with only 2.5 percent in the form of freshwater. Of this 2.5 percent, 70 percent is locked up in ice or permanent snow cover in mountainous regions and the Antarctic and Arctic regions, while 29.7 percent is stored below the ground (groundwater). Surface water, including rivers and lakes, comprise the remaining 0.3 percent of freshwater resources available.[1] A water molecule is made up of two hydrogen atoms bonded to a single oxygen atom. The connection between atoms is through covalent bonding: the sharing of an electron from each atom to give a stable pair. In the water molecule structure,

Urban Water Security, First Edition. Robert C. Brears.
© 2017 John Wiley & Sons, Ltd. Published 2017 by John Wiley & Sons, Ltd.

the hydrogen atoms are not arranged around the oxygen atom in a straight line; instead there is an angle of approximately 105° between the hydrogen atoms.[2] The hydrogen atoms are positive and so do not attract one another, while the oxygen atom has two non-bonding electron pairs that repulse the two hydrogen atoms.

Water molecules are described as bipolar because there is a positive and negative side of the molecule. This enables water molecules to bond with one another; this is known as hydrogen bonding. In hydrogen bonding, the positive side of the water molecule (the hydrogen side) is attracted to the negative side (the oxygen side) of another water molecule, and a weak hydrogen bond is formed.[3] The hydrogen bonding of water molecules is responsible for a number of water's properties. For instance, based on water's molecular weight (MW = 20), water should evaporate and become a gas at room temperature, given that $CO_2(MW = 44)$, $O_2(MW = 32)$, $CO(MW = 28)$, $N_2(MW = 28)$, $CH_4(MW = 18)$ and $H_2(MW = 2)$ are all gases at room temperature. The reason why water does not evaporate at room temperature is due to water's high specific heat capacity (a temperature increase is effectively an increase in the motion of molecules and atoms comprising the substance). When water is heated, it causes a movement of water molecules – breaking of the hydrogen bonds. However, due to water's cohesiveness, water molecules have a high resistance to increasing their motion. Therefore, it requires a lot of energy to break the hydrogen bonds. As such, water does not evaporate easily. This high heat capacity means water is resistant to radical swings in temperature which is taken advantage of by organisms. Other properties of water include adhesiveness – water molecules are attracted to other substances such as chemicals, minerals and nutrients; solvency – water is a universal solvent as it can dissolve more substances than any other liquid on Earth and uniqueness – water is unique as its solid form (ice) is less dense than liquid water, and it can change from ice to water vapour without first becoming a liquid.[4]

1.2 Hydrological cycle

The hydrological cycle is the continuous movement of water in all its phases: liquid (precipitation), solid (ice) and gaseous (evaporation) forms. Because water is indestructible, the total quantity of water in the cycle does not diminish as water changes from vapour to liquid or solid and back again. In this cycle, evaporation from oceans (505 000 cubic kilometres) exceeds the 458 000 cubic kilometres of precipitation that falls on them. Meanwhile, 119 000 cubic kilometres of precipitation falls on land, which comprises one third of the Earth's surface, and 72 000 cubic kilometres returns through evaporation to the atmosphere. The difference (47 000 cubic kilometres) is either ground or surface water that eventually returns to the ocean.[5] The average amount of time a water molecule remains in a particular part of the hydrological cycle is known as its residence time. Streams and rivers usually have residence times of only days or months, while lakes and inland seas have residence times of years to decades. In comparison, oceans and groundwater systems have residence times of 3000–5000 years (Table 1.1).[6]

Table 1.1 Principal residence times of the global water stores

Compartment	Volume (1000 cubic kilometres)	Percent	Mean residence time (years)
Oceans	1 370 000	93.943	3000
Groundwater	60 000	4.114	5000
Actively exchanging groundwater	4 000	0.274	300
Glaciers and ice caps	24 000	1.646	8600
Lakes/inland seas	230	0.016	10
Soil water	82	0.006	1
Atmospheric vapour	14	0.001	0.027
Rivers	1.2	0.0001	0.032

CLOSS, G., DOWNES, B. J. & BOULTON, A. J. 2004. *Freshwater Ecology: A Scientific Introduction.* Malden, MA: Wiley-Blackwell

The hydrological cycle contains four key components: precipitation, runoff, evaporation and groundwater storage.

1.2.1 Precipitation

Atmospheric vapour, which results in precipitation in both liquid (rainfall) and solid (snow) forms, accounts for less than 0.001 percent of the world's total water; however, due to its low residence times in the atmosphere, it is one of the main drivers of the hydrological cycle.[7]

Precipitation occurs when a body of moist air is cooled sufficiently for it to become saturated. Air can be cooled by a meeting of air masses of differing temperatures or by coming into contact with cold objects such as land surfaces. However, the most important cooling mechanism is the uplifting of air: as warm air rises, its pressure decreases while it expands and cools.[8] This cooling reduces the air's ability to hold water vapour and condensation forms. Condensation is composed of minute particles floating in the atmosphere, providing a surface for water vapour to condense into liquid water. Water or ice droplets formed around condensation particles are usually too small to fall directly to the ground as precipitation due to the upwards draught within the cloud being greater than the gravitational forces pulling the droplets down. In order to have a large enough mass to fall, raindrops grow through collision and coalescence. In this process, raindrops collide and join together (coalesce) to form larger droplets that collide with many other raindrops before falling towards the surface as precipitation. Whether precipitation is rain or snow depends on the warmth of the clouds. In warm clouds temperatures are above freezing point, and water droplets grow through collision (the coalescence process) to form rain. In cold clouds temperatures are below freezing point. These clouds contain ice crystals and supercooled water that is liquid water chilled below its freezing point without it becoming solid. In these clouds precipitation is in the form of snow.[9]

There are three types of precipitation: frontal and cyclonic, convectional and orographic precipitation. Frontal precipitation occurs in the narrow boundaries or fronts between air masses of large-scale weather systems. In this system, warm moist air is forced to rise up and over a wedge of colder, dense air. There are both warm and cold fronts each distinguished by the resulting precipitation: cold fronts have steep frontal surface slopes causing rapid lifting of warm air, resulting in heavy rain over a short duration, while warm frontal surfaces are much less steep, causing gradual lifting and cooling of air, leading to less intense rainfall but over a longer duration.[10] In cyclonic systems, there is a convergence and rotation of uplifting air. In the northern hemisphere, cyclonic systems rotate anticlockwise and in the southern hemisphere clockwise. Above and below the tropics in the northern and southern hemispheres, cyclonic systems usually have a weak vertical motion, resulting in moderate rain intensities for long durations, while in the tropics, because of greater heating of the air, there is more intense precipitation but of a shorter duration.[11] Convectional precipitation happens when the ground surface of a landmass causes warming of the air: as the warm air rises, it cools down and condenses, leading to localised, intense precipitation of a short duration. As this type of precipitation is dependent on the heat of the landmass, it is most common over warm continental interiors such as Australia and the United States. However, this type of precipitation does occur over tropical oceans with slow-moving convective systems producing significant amounts of rainfall. It is common for clusters of thunderstorm cells to be embedded inside convective systems, which commonly leads to flooding events.[12] Orographic precipitation is the result of moist air passing over land barriers such as mountain ranges or islands in the ocean. The South Island of New Zealand is an example of orographic precipitation: the warm moist air off the Tasman Sea reaches the West Coast of the South Island, and as it starts to lift over the Southern Alps, the warm moist air cools and condenses, producing significant rainfall on the West Coast, while on the leeward side the air descends and warms up resulting in low levels of cloud and rainfall.[13]

1.2.2 Runoff

Runoff, or streamflow, is the gravitational movement of water in channels. A channel can be of any size ranging from small channels in soils with widths in the millimetres to channels of rivers. The unit of measurement for runoff is the cumec, with one cumec being one cubic metre of water per second. Streamflows react to rainfall events immediately indicating that part of the rainfall takes a rapid route to the stream channel. This is known as quick flow, while base flow is the continuity of flow even during periods of dry weather.[14] Precipitation can arrive in stream channels through four ways: direct precipitation, overland flow, throughflow and groundwater flow. Direct precipitation comprises only a small amount of streamflow as channels usually occupy only a small percentage of the surrounding area; therefore, it is only during prolonged storms or precipitation events that direct precipitation contributes significantly to streamflow. Overland flow is water that instead of infiltrating soil flows over the ground surface into stream channels during periods of high-intensity rainfall. Overland flows usually occur on moderate to steep slopes in arid and semi-arid areas as these areas lack vegetation and so have dry, compact soil.[15] Throughflow is all the water that infiltrates the soil surface and moves laterally towards a stream channel. This type of flow occurs during periods of prolonged or heavy rainfall when water enters the upper part of the soil profile more rapidly than it can drain vertically. Finally, groundwater flow is water that has percolated through the soil layer to the underlying groundwater and from there into the stream channel.[16]

1.2.3   Evaporation

Evaporation is the transferral of liquid water into a gaseous state followed by its diffusion into the atmosphere. The presence or lack of water at the surface provides the distinctions in definitions for evaporation.[17] For instance, open water evaporation (E) occurs above a body of water such as a lake, stream or ocean. Potential evaporation (PE) is evaporation that would occur if the water supply was unrestricted, while actual evaporation (AE) is the quantity of water that is actually removed from a surface due to evaporation.

Evaporation over a land surface occurs two ways, either as actual evaporation from the soil or transpiration from plants. Transpiration occurs as part of photosynthesis and respiration and is controlled by the plant leaf's stomata opening and closing.[18] The main source of energy for evaporation is the sun. The term used to describe the amount of energy received from the sun at the surface is net radiation ($Q*$), and its calculation is

$$Q* = QS \pm QL \pm QG$$

where QS is sensible heat, the heat we feel as warmth; QL is latent heat and is the heat absorbed or released during water's phase change from ice to liquid water or liquid water to water vapour (there is a negative flux (when energy is absorbed) when water moves from liquid to gas and a positive flux when gas is converted to liquid) and QG is solid heat flux and is the heat released from the soil that has previously been stored within the soil.[19]

1.2.4   Groundwater

Below the Earth's surface, water can be divided into two zones – unsaturated and saturated. In the unsaturated zone, water is referred to as soil water and occurs above the water table, while the saturated zone is referred to as groundwater and occurs beneath the water table. In the unsaturated zone, the majority of water is held in soil that is composed of solid particles (minerals and organic matter) and air. The infiltration rate is used to determine how much water enters the soil over a specific period of time. The rate is dependent on the current water content of the soil and the soil's ability to transmit water. For instance, soil that has high moisture content will have a low infiltration rate because water has already filled voids between the soil's solid particles.[20]

Once water has infiltrated the unsaturated zone, it percolates down through the water table to become groundwater. Groundwater can be found at depths of 750 metres below the surface. It is estimated that the volume stored as groundwater is equivalent to a layer of water approximately 55 metres deep spread over the entire Earth's landmass.[21] Most groundwater is in motion; however, unlike stream and river flows, groundwater moves extremely slow at rates of centimetres per day or metres per year with the actual rate dependent on the nature of the rock and sediment it passes through. Porosity is the percentage of the total volume of a body of rock that contains open spaces (pores). Therefore, porosity determines the amount of water rocks can contain, while porosity in sediments is dependent on the size and shape of the rock particles it contains and the compactness of their arrangement.[22] Meanwhile, permeability is the measure of how easily a solid allows fluid to pass through. Rocks with a very low porosity are likely to have low permeability; however, rocks with high porosity does not mean they have high permeability. Instead, it is the size of the pores, how well they are connected and how straight the path is for water to flow through the porous material that determines the permeability of a rock or sediment.[23]

An aquifer is a body of highly permeable rock, typically gravel and sand, that can store water and yield sufficient quantities to supply wells, while an aquitard is a geological formation that transmits water at a much slower rate (aquitards are usually defined as a formation that confines the flow over an aquifer, while the term aquifuge is sometimes used to define a completely impermeable rock formation).[24] There are two types of aquifers: confined and unconfined. A confined aquifer has a boundary (aquitard) above and below it that constricts the water into a confined area. Geological formations are usually the most common form of confined aquifers because they often occur as layers, and so the flow of water is restricted vertically but not horizontally.[25] Water in confined aquifers is normally under pressure: when it is intersected by a borehole, it will rise up higher than the restrictive boundary. If the water rises to the surface, then it is known as an artesian well. Unconfined aquifers have no boundaries above, and so the water table is free to rise and fall depending on the amount of water in the aquifer.

The movement of groundwater can be described by Darcy's law: Henry Darcy was a nineteenth-century French engineer who conducted observations on the characteristics of water flowing through sand. Darcy observed that the rate of flow through a porous medium was proportional to the hydraulic gradient. The most common formula for Darcy's law is

$$Q = -k_{sat} \times A \times \frac{dh}{dx}$$

The discharge (Q) from an aquifer equals the saturated hydraulic conductivity (k_{sat}) multiplied by the cross-sectional area (A) multiplied by the hydraulic gradient

(dh/dx). The negative sign is based on the fact that a fall in gradient is negative.[26] The h term in the hydraulic gradient includes both the elevation and pressure head.

1.2.5 How old is water?

Determining the age of water is important for managing water resources as the age provides an indication of how quickly contaminated water can move towards an extraction zone and how long ago the contamination occurred. Because Darcy's law cannot be used to determine the time it takes for water to reach a certain position, scientists instead conduct chemical analyses of dissolved substances in water to estimate its age. Carbon dating is common for testing the age of groundwater; however, it is problematic for young groundwater because it is only accurate if the sample is more than thousand years old.[27] When testing old groundwater, carbon dating involves the analysis of the rate of decay of ^{14}C in dissolved organic carbon. For younger groundwater, chemical dating of water involves determining the concentrations of material that humans have polluted the atmosphere with as these substances are dissolved in precipitation. The concentrations of these substances provide an estimate on the average age of the groundwater tested. Tritium is a radioactive isotope of hydrogen and was added to the atmosphere in large quantities as a result of hydrogen bomb tests in the 1960s and 1970s. Tritium concentrations in the atmosphere peaked in 1963 and have since declined to background levels.[28] This particular radioactive isotope has a half-life of 12.3 years. Chlorofluorocarbon (CFC) compounds were commonly used in aerosols and refrigeration from the 1940s until they were banned in the 1990s. There are two CFC compounds: CFC-11 which has slowly declined since 1993 and CFC-12 which is still increasing but at a slower rate than before 1990. Sulphur hexafluoride is used for cooling and insulation mainly in electronics.

Another method for dating groundwater is analysing the ratio of the two isotopes of oxygen and/or the two isotopes of hydrogen found in water molecules. When water in the atmosphere condenses to form rain, there is a preferential concentration of heavy isotopes of hydrogen and oxygen in the water molecules.[29] The heavy isotope of hydrogen is known as deuterium, and the heavy isotope of oxygen is ^{18}O, and the colder the temperature at the time of condensation, the more enriched in deuterium and ^{18}O the water sample is. Therefore, in climates with distinct seasons, the amount of deuterium and ^{18}O will vary with each season, and so if the groundwater shows variations in deuterium and/or ^{18}O, then it comprises relatively new rainfall. If there is little variation in deuterium and/or ^{18}O, it indicates that there has been mixing of rainfall from both past summers and winters and therefore it is older.[30]

1.3 Natural variations to water quantity

There are two types of natural variations to water quantity: floods and droughts.

1.3.1   Floods

Floods occur when precipitation and runoff exceed the capacity of the river channel to carry the increased discharge. Flood frequencies are used when planning land use and infrastructure design and are calculated based on the history of a river, that is, how often it has flooded in the past and what the historical extremes of high precipitation are. Flood frequencies are expressed as a recurrence interval – the probability a particular flood will occur in a given year, for example, a hundred-year flood means there is a one in a hundred chance of it occurring in that particular year.[31] Recurrence intervals are calculated using models that incorporate probable maximum precipitation (PMP) and probable maximum flood (PMF) calculations. The PMP is the finite limit for precipitation from a single storm event – the maximum depth (amount) of precipitation that is reasonably possible during a single storm event. Flood events have maximum extremes, and the PMF is the maximum surface water flow in a drainage area that could be expected from a PMP event.[32] Floods can cause significant damage to buildings and properties with water washing away soils and crops, depositing sediments on land and property and be potentially fatal to humans and animals. Services are usually designed to resist floods or be serviceable against the following probabilities: important roads are designed to withstand a hundred-year floods, that is, a 1 percent chance of being overtopped in any given year; general roads and buildings are designed to withstand 50-year floods and less important roads, 20-year floods and storm water drains and pipes can be designed to withstand anything from a 2- to 20-year recurrence interval depending on the consequences over overtopping.[33]

1.3.2   Droughts

A drought is a period of unusually dry weather that persists over a long enough period of time to cause crop damage and/or water supply shortages. There are four different ways a drought can be defined. Meteorological droughts are a measured departure of precipitation from normal levels. Agricultural droughts refer to situations in which the amount of moisture in the soil no longer meets the needs of a particular crop. Hydrological droughts occur when surface and groundwater supplies are below normal levels. Socioeconomic droughts occur when physical water shortages begin to affect people.[34,35] Droughts have varying levels of severity and return periods ranging from minor droughts that have a return period of 3–4 years, with slowing of growth in crops and pastures, to exceptional droughts with a return period of over 50 years with widespread crop and pasture loss and shortages of water in reservoirs (Table 1.2).

Both the onset and end of droughts can be predicted by meteorologists observing precipitation patterns, soil moisture and streamflow data. To do this, meteorologists use a variety of indices that show deficits in precipitation over a period of time. One common tool is the Standardised Precipitation Index (SPI), which is a drought index based on the probability of an observed precipitation deficit occurring over a period of time ranging from 1 to 36 months. This variable time-scale allows the index to describe drought conditions important for a range of

Table 1.2 Drought severity classification

Drought severity	Return period (years)	Description of possible impacts
Minor	3–4	Going into drought: short-term dryness slowing growth of crops or pastures Coming out of drought: some lingering water deficits, pastures and crops not fully recovered
Moderate	5–9	Some damage to crops or pastures, streams, reservoirs or wells low, some water shortages, developing or imminent voluntary water restrictions requested
Severe	10–17	Crop or pasture losses likely, water shortages common, water restrictions imposed
Extreme	18–43	Major crop and pasture losses, widespread water shortages or restrictions
Exceptional	44+	Exceptional and widespread crop and pasture losses, shortages of water in reservoirs, streams and wells, creating water emergencies

SMITH, K. 2013. *Environmental Hazards: Assessing Risk and Reducing Disaster*. Hoboken, NJ: Taylor & Francis

Table 1.3 Palmer Drought Severity Index

Index	Description
4.0 or more	Extremely wet
3.0 to 3.99	Very wet
2.0 to 2.99	Moderately wet
1.0 to 1.99	Slightly wet
0.5 to 0.99	Incipient wet spell
0.49 to −0.49	Near normal
−0.5 to −0.99	Incipient dry spell
−1.0 to −1.99	Mild drought
−2.0 to −2.99	Moderate drought
−3.0 to −3.99	Severe drought
−4.0 or less	Extreme drought

CENTER, N. D. M. 2011. *Comparison of major drought indices: Palmer Drought Severity Index* [Online]. Available: http://www.drought.unl.edu/Planning/Monitoring/ComparisonofIndicesIntro/PDSI.aspx (accessed 10 May 2016)

meteorological, agricultural and hydrological applications. For example, soil moisture responds to a precipitation deficit immediately, while groundwater recharge and reservoir levels respond to precipitation deficits over many months. When describing the severity of droughts, the common index used is the Palmer Drought Severity Index. This index is a soil moisture algorithm that includes water storage and evapotranspiration levels with a scale ranging from ≥4.0 (extremely wet) to ≤−4.0 (extreme drought) (Table 1.3).

1.4   Natural variations to water quality

Natural processes, including temperature, dissolved oxygen, pH, dissolved and suspended solids, turbidity, minerals, salinity, inorganic and organic chemicals and nutrients, affect the quality of water resources, specifically those discussed in the following text.

1.4.1   Temperature

Numerous physical, biological and chemical characteristics of water bodies are dependent on temperature. For instance, temperature is an important signal for spawning and migration. Sudden changes in temperature can be deadly for many species, and this usually occurs when deep, cold reservoir water is released into warm waterways.[36,37] Temperature and dissolved oxygen are interdependent with warmer water holding less dissolved oxygen than colder water.

1.4.2   Dissolved oxygen

The presence or absence of dissolved oxygen in an aquatic ecosystem is one of the main determinants of whether organisms can live in that environment or not. Habitats that have a presence of oxygen are aerobic, while environments lacking dissolved oxygen are anaerobic.[38] Dissolved oxygen levels are an indicator of water quality, with high concentration levels indicating high water quality. Oxygen however is only slightly soluble in water, and so there is high competition among aquatic organisms including bacteria for it. Dissolved oxygen is important for aquatic plants and animals as it allows species to breathe.[39] When dissolved oxygen levels decrease below 5 milligram per litre, most sensitive organisms such as fish become stressed. If dissolved oxygen levels reach 1 milligram per litre, most species will not survive for more than a few hours.[40]

1.4.3   pH

The pH (p, power; H, hydrogen) level of a solution indicates its basicity or acidity, and it is defined as the negative logarithm of the hydrogen proton. Solutions with a pH less than 7 are said to be acidic, and those with a pH greater than 7 are basic or alkaline. Because the scale used to measure pH is logarithmic, each number represents a 10-fold change in the proton activity in a solution. Therefore, water with a pH of 4 is 10 times more acidic than that with a pH of 5.[41] Different water bodies have differing pH levels, for instance, bogs and wetlands have acidic conditions with pH levels between 4 and 7, while water in rivers and lakes usually have pH levels between 4 and 9. Fish in water bodies usually have a narrow range of pH preference which varies greatly with specie. If the pH level of a water body changes to a level outside a fish's preferred level, it can cause physical damage to skin, gills and eyes and eventually be fatal.[42]

1.4.4   Dissolved and suspended solids

As water passes through the soil column or over a surface, it dissolves substances attached to the soil particles. Water also dissolves particles in the air as it passes through the atmosphere in the form of rain. The amount of dissolved substances in a water sample is known as the total dissolved solids (TDS), and the higher the TDS, the more contaminated the water body is.[43] TDS can also be used to estimate the conductivity of water. Conductivity is the amount of electricity that can be conducted by water, and the more the ions present, the higher the conductivity. Conductivity is correlated roughly to productivity because high-nutrient water has high conductivity.[44]

The measuring of total suspended solids (TSS) is another key measure of water quality. Rivers and streams carry suspended sediment as part of the natural erosion and sediment transport process in which sediment is deposited/picked up whenever river velocity decreases/increases. Soil particles are usually naturally carried as suspended load in water bodies. However, events such as landslides remove natural vegetation exposing bare soils. This can lead to excessive suspended loads in water bodies, increasing turbidity and decreasing water clarity.[45] When sediment enters waterways and becomes suspended in the water body, it can severely damage the wildlife inhabiting the waterway. For instance, suspended sediment abrades and damages fish gills, increasing the risk of infection, disease and death. This leads to the loss of sediment-sensitive fish species. Suspended sediment also reduces sight distance for fish, reducing feeding efficiency. It also blocks light from entering the water, reducing photosynthesis in plants, leading to a reduction in aquatic food for many species. Deposited sediments also affect aquatic wildlife in waterways. For instance, it physically smothers benthic aquatic insect communities, which in turn reduces the amount of food available for species higher up the food chain. Deposited sediments also cover and destroy spawning grounds reducing fish populations. It also smothers fish eggs reducing their survival rates.[46]

1.4.5   Turbidity

Turbidity is the measure of clarity in water and is dependent on the amount of suspended matter in the water that reduces transmission of light. It is caused by suspended matter, including clay, silt and organic material, in the water creating cloudiness. High turbidity levels indicate that there are problems in the water body as turbidity blocks out sunlight needed for aquatic vegetation, impacting on the health of the aquatic ecosystem. Turbidity can also create water quality problems with toxic chemicals attaching themselves to suspended particles in water bodies.[47] Because turbidity is a measure of the cloudiness of water, TSS and turbidity are directly related.[48]

1.4.6   Minerals

As water moves through the terrestrial system, materials containing minerals are dissolved or weathered from the land. Chemical weathering involves the dissolving of materials, while mechanical weathering reduces particles of matter to sizes that may be dissolved at a later stage. The total concentration of dissolved solids carrying minerals is inversely dependent on the amount of runoff – the greater the runoff, the less time taken for water to dissolve the ions.[49] The minerals that enter water bodies through chemical and mechanical weathering are important for plant and animal life as minerals are needed to control chemical reactions. The main minerals required in human diets include calcium, magnesium, phosphorus and potassium.

1.4.7   Salinity

Surface water runoff and groundwater percolation from precipitation and irrigation can cause salts to leach, that is, dissolve from the soil and contaminate surface and groundwater supplies. The term salinity is used to describe the presence of excess salts in water and is harmful to certain plants, aquatic species and humans. In humans, high salt levels in water can lead to increased blood pressure, while in plants saline soils harm plants by pulling moisture out of the root system reducing the uptake of water and fertiliser.[50]

1.4.8   Inorganic and organic chemicals

Inorganic chemicals are any chemicals that do not contain carbon. In low quantities, metals such as calcium, zinc and iron are healthy for the human body, while copper, lead, mercury and arsenic can be toxic or poisonous. Organic chemicals contain carbon and hydrogen and can be classified as natural or synthetic with natural organic chemicals extracted from sugars, carbohydrates, amino acids and proteins. Synthetic chemicals are mass-produced and persist in the environment for long durations. This is because natural enzymes are unable to break down their complex compounds. In addition, many synthetic chemicals are carcinogens and can be divided into two categories: volatile organic chemicals that are lightweight and dispersed through the air and non-volatile chemicals that are heavy and settle at the bottom of rivers and lakes into sediments.[51] Pesticides are synthetic organic chemicals and include insecticides for killing insects and herbicides for killing plants and weeds. They are designed to be applied to a target area to control a specific pest and then degrade; however, pesticides frequently contaminate ground and surface water supplies.

1.4.9   Nutrients: Nitrogen and phosphorus

The most common form of nitrogen in the biosphere is nitrogen gas, which comprises 78 percent of the atmosphere. In water, nitrogen is dissolved as a gas and is less soluble than oxygen in water; however, despite being less soluble than oxygen, its higher atmospheric concentration means its dissolved concentration is similar to oxygen.[52] Nitrogen is an important nutrient in water quality and exists as five main types: proteins, amino acids and urea, nitrite, nitrate and ammonia. However, when excessive nitrogen enters a water body, an oxidation process called nitrification occurs. There are two associated problems with nitrification occurring in natural water bodies: first, oxygen demand increases, and second, nitrification is highly toxic resulting in fish populations dying or migrating.[53]

Phosphorus is required in large amounts by plants and is one of the most important nutrients for plant growth in aquatic ecosystems. The main sources of phosphorus are phosphorus-bearing minerals such as iron, aluminium and calcium phosphates that occur in low concentrations in soils.[54] Therefore, in natural unpolluted waterways the primary source of phosphorus is watershed soils and bottom sediment. Rock phosphate is mined and processed to form a highly soluble calcium phosphate compound for use as agricultural, domestic and industrial phosphates. The main concern of excessive nitrogen and phosphorus in water bodies is eutrophication. Nitrogen enhances the growth of not only agricultural plants but also aquatic plants including algae, leading to the overproduction of plant matter in lakes, rivers and streams.[55] The negative impact of excessive aquatic plant and algae growth in water bodies is the depletion of dissolved oxygen caused by the decomposition of dead vegetative matter, resulting in the subsequent decline in aquatic species numbers and water body health.[56]

1.5   Impacts of urbanisation on water resources

Urbanisation into river basins can lead to pollution of rivers, lakes and wetlands from point and non-point source pollution. In addition, urbanisation impacts aquatic ecosystems, while impervious surfaces lower water quality and groundwater levels.

1.5.1   Point source pollution

Point source pollution is the contamination of a water body through a pipe or other clearly identified location. This type of pollution is easily measured and impacts assessed. The most common sources of point source pollutants are factories, wastewater treatment plants, landfills and underground and aboveground storage tanks holding fuel, solvents and other industrial liquids.[57] Industrial-related wastewater can contain a number of pollutants including microbiological contaminants, chemicals from industrial activities including solvents, organic and inorganic chemicals and heavy metals. In addition, point source pollution can include suspended matter and the discharge of warm water into cooler waterways.[58]

1.5.2   Non-point source pollution

Non-point source pollution is generated from numerous sources, and therefore its origin is difficult to identify. Fertiliser runoff from gardens and agricultural activities is a common source of non-point source pollution. Nitrate is commonly used as a fertiliser because it is highly soluble and therefore easily taken up by plant's root systems. However, because it is highly soluble, it is easily flushed through the soil into rivers and streams resulting in eutrophication of waterways.[59] In urban areas precipitation events flush large amounts of pollutants including heavy metals and oils into water bodies.[60] Groundwater is a significant source of drinking water; however it is one of the most neglected sources of water due to its low visibility. Pollution of groundwater is a serious issue, even if the source of the pollution is removed, due to groundwater's high residency times – groundwater can remain contaminated for hundreds of years. There are many sources of groundwater contamination that includes industrial wastes, septic tanks, landfills, agriculture, municipal landfills, mining and petroleum products and saltwater intrusion.[61]

Runoff from roads and highways contain numerous types of contaminants. Sediments in runoff from roads are usually due to the clearing of land near roads for construction. When sedimentation enters nearby waterways, it reduces the amount of light that can penetrate the water, affecting photosynthesis rates of aquatic plants. It can also damage fish gills causing disease and death. When sediments settle on the beds of waterways, it smothers spawning grounds further reducing fish populations. Fertilisers are commonly applied near roads for plant vegetation. Runoff containing fertiliser can lead to excessive algal growth in waterways nearby; eventually decreasing the water's dissolved oxygen content and killing aquatic life. Heavy metals are commonly found in waterways near transport routes. Heavy metals often adhere to sediment, degrading water quality and harming aquatic wildlife by interfering with the processes of photosynthesis, respiration, growth and reproduction.[62]

1.5.3   Damage to aquatic ecosystems

The impacts of urbanisation on aquatic ecosystems include loss of wetland and riparian buffers. Stream ecosystems are dependent on extensive freshwater wetlands, floodplains, riparian buffers, springs and flood channels – all of which are lost during urbanisation. Hard (impervious) surfaces and accompanying storm water systems can cause lower base flows in streams and faster runoff during storms. Water running off impervious surfaces often has a higher temperature than naturally flowing water due to its higher residence time on hard surfaces. Meanwhile intensive urbanisation can raise stream water temperatures by 5–10 degrees Celsius due to the loss of shading from riparian vegetation and lower water levels. Urbanisation usually leads to a shift in energy sources. In natural streams the aquatic ecosystem is driven by an energy source comprising decomposing vegetation, woody debris and falling insects, all of which is lost through urbanisation. In urban waterways, reduced tree canopies in addition to nutrient accumulation results in an increase in aquatic plants and algae, significantly lowering the health of the overall aquatic ecosystem. There is also a reduction in biological diversity with urban waterways only supporting a fraction of the fish and aquatic invertebrates that would exist in an undeveloped waterway.[63]

1.5.4   Impervious surfaces modifying hydrological cycles

The hydrological cycle is the continuous movement of water between land, water bodies and atmosphere. When precipitation reaches the surface, some evaporates, some percolates through the soil becoming groundwater and the remainder becomes surface water. Impervious cover (hard surfaces that do not allow water to penetrate the soil, for example, streets, driveways and rooftops), however, alters the natural amount of water that takes each path of the hydrological cycle. In urban areas, impervious cover and urban drainage systems increase the volume and velocity of surface runoff into waterways. It has been estimated that in areas of natural groundcover, a quarter of rainfall infiltrates the soil and becomes groundwater while only 10 percent ends up as surface runoff. As impervious cover increases with urbanisation, 20 percent of rainfall becomes surface water, while in highly urbanised areas, 55 percent of rainfall becomes surface water.[64] This increased surface water causes severe erosion of stream and riverbanks and transportation of sediment, clogging stream channels and damaging natural habitats. In addition, there is a larger volume and faster discharge of surface water during storm events compared to natural lands, resulting in more flooding and habitat damage. Because of the increased surface runoff in urbanised areas, there is greater risk of flooding, and so many waterways become drainage channels that are frequently lined with rocks and concrete. The result of this is loss of riparian vegetation and habitats for aquatic wildlife.

1.5.5   Impervious surfaces lowering water quality

Increased impervious cover also results in lower water quality in waterways because pollutants, collected on impervious surfaces, are frequently washed into streams, rivers and lakes. In catchments with less than 10 percent impervious cover, waterways remain healthy; however, above 10 percent stream degradation is frequent and includes excessive stream erosion, reduced groundwater recharge, increased size and frequency of floods, loss of riparian vegetation, increased contaminants in water and decrease in stream biodiversity. In addition, contaminated surface water containing pollutants including nitrogen compounds, dissolved organic carbon, synthetic organic compounds and petroleum products can infiltrate the surface, severely degrading groundwater quality.[65]

1.5.6   Impervious surfaces affecting groundwater recharge

Many urban areas are dependent on groundwater supplies for reticulated public water supplies and domestic and industrial use. Urbanisation can affect the groundwater system by changing the patterns and rates of aquifer recharge.

In urban areas where abstraction of groundwater is heavy and exceeds the rate of local recharge, aquifer levels may continue to decline over decades resulting in deepening of wells and declining water table levels. This can lead to the intrusion of saline water and ground subsidence, resulting in physical damage to buildings and underground engineering structures and services such as tunnels and sewers.[66]

While impervious surfaces can reduce normal soil infiltration of water paths, car parks and other low-permeability surfaces unconnected to storm water drains can recharge urban groundwater. In addition, water mains can leak around 20–25 percent of water carried and lead to further recharge of groundwater as can excess irrigation of gardens and parks.[67] Increased groundwater recharge can cause hydrostatic uplifting of the surface resulting in damage to infrastructure. It can also lead to rising water tables inundating subsurface infrastructures such as building foundations and basements.[68]

1.6   Water and wastewater treatment processes

When it comes to consumption of drinking water, there are two types of drinking water standards: primary and secondary. Primary standards are designed to protect public health by establishing maximum permissible levels of potentially harmful substances in water. Secondary standards apply to aesthetic aspects of drinking water that do not pose a risk to health (such as colour and odour).[69] Water utilities use both natural and chemical processes to ensure drinking water is free of contamination to meet primary standards. Drinking water purification comprises four stages: sedimentation, coagulation and flocculation, filtration and disinfection. Impurities in drinking water are either dissolved or suspended solids. Under the process of sedimentation, water under quiescent conditions has minimal flow velocities and turbulence, and so particles denser than water can settle out at the bottom of a tank in the form of sludge.[70] Not all suspended particles are removed during the process of sedimentation with very small turbidity-causing particles called colloids remaining. In the coagulation stage, chemicals called coagulants are mixed into the water causing particles in the water to stick together and form larger and heavier particles called flocs. After the chemicals are added, the water is slowly stirred – a process called flocculation – and this increases the sticking of particles to one another. The combined process of applying chemicals and then stirring of the water is known as coagulation.[71] After coagulation around 5 percent of suspended soils can remain as non-settling floc particles. Filtration involves the removal of suspended particles from water by passing it through filter beds of porous granular material such as sand. As water passes through the filter bed, the suspended particles become trapped within the pore spaces. Because pores eventually become blocked, backwashing occurs at times – a process involving clean filtered water being forced back up through the filter carrying away the accumulated particles. Coagulation, sedimentation and filtration remove nearly all microorganisms and suspended sediments from the water. However, it is usually not enough to remove completely all pathogen bacteria and viruses present in the water. To achieve this, the final treatment of water is disinfection involving most commonly chlorination but also ozone and ultraviolet radiation treatment.[72]

To reduce the potential for waterborne disease and damage to ecosystem health, wastewater is treated before it is returned to the natural environment. Wastewater treatment comprises three stages: primary, secondary and tertiary. Primary treatment involves the removal of suspended solid material from wastewater. Floating material such as wood, paper and oil are removed first; otherwise it will block the filters and pipes. Wastewater is then pushed into a grit chamber where sand and small stones settle to the bottom of the chamber. This process is common in areas with combined storm water drainage and sewer systems because in these systems sand and gravel often wash into sewers after storm events. Following this, wastewater proceeds into primary settling tanks where suspended solids settle as sludge for removal.[73] The secondary treatment process involves the removal of oxygen-demanding organic matter. This involves two types of processes; trickling filters and activated sludge systems. Wastewater is passed through trickling filters which are beds filled with coarse material comprising rocks and gravel. As the wastewater passes through the beds, a microbial filter develops on the surface of the rocks and gravel trapping oxygen-demanding organic matter as the wastewater passes through. In the activated sludge system stage, the effluent is constantly agitated and aerated so that sludge forms. This sludge contains aerobic organisms that digest any remaining organic material. By now the wastewater contains only between 5 and 20 percent of its original organic matter and can be safely discharged into waterways. However, nitrates and phosphates still remain and require tertiary treatment. Nitrogen is removed through the biological oxidation of nitrogen from ammonia to nitrate (nitrification) followed by denitrification, which is the reduction of nitrate to nitrogen gas. From which, nitrogen gas is released to the atmosphere (removed from the water). Meanwhile, phosphorus can be removed biologically using a process called enhanced biological phosphorus removal that involves special bacteria called polyphosphate-accumulating organisms that accumulate large quantities of phosphorus within their cells. These organisms can then be separated from the treated water to form a bio-solid that can be used as a fertiliser.[74]

1.6.1 Ensuring drinking water safety

The World Health Organization's (WHO) *Guidelines for Drinking-water Quality* provides a guidance on how to develop and implement risk management strategies to ensure safety of drinking water supplies. The guidelines outline a preventative management framework for safe drinking water that comprises five key components:

    1 *Establish health-based targets*: Health-based targets should be established by a high-level authority responsible for health in consultation with water suppliers, affected communities, etc. The targets should take into account the overall public

health situation and contribution of drinking water quality to disease from water-borne microbes and chemicals as part of an overall water and health policy.

2 *Conduct system assessments*: System assessments determine whether the drinking water supply can deliver water that meets health-based targets: assessment of the drinking water supply system is applicable for large utilities, small community supplies and individual domestic supplies. Assessments can be of existing infrastructure, plans for new supplies or upgrades of existing supplies. As drinking water quality varies throughout the system, assessments should aim to determine if the final quality of water delivered to consumers routinely meets established health-based targets.

3 *Conduct operational monitoring*: Operational monitoring is the conducting of planned observations or measurements to assess whether control measures that ensure drinking water quality are operating properly. Usually, operational monitoring involves simple and rapid tests, for example, turbidity rather than complex microbial or chemical tests, which are generally conducted as part of the validation and verification activities.

4 *Implement management plans*: Management plans document system assessment and operational monitoring and verification plans describe actions in both normal operation and during incidents where loss of control of the system may occur. The management plan should also outline procedures and programmes required to ensure optimal operation of the drinking water system.

5 *Conduct independent surveillance*: The surveillance agency is responsible for independent and periodic review of all aspects of safety, while the water supplier is responsible for regular quality control, operational monitoring and ensuring good operating practices. Surveillance contributes to the protection of public health by promoting the improvement of quality, quantity, accessibility, coverage, affordability and continuity of drinking water supplies. Surveillance requires a systematic programme of surveys that cover the whole drinking water system, including sources, activities in catchments, transmission infrastructure, treatment plants, storage reservoirs and distribution systems.[75]

Notes

1. UN-WATER. 2013. *Statistics* [Online]. Available: http://www.unwater.org/statistics/statistics-detail/en/c/211803/ (accessed 2 June 2016).
2. DAVIE, T. 2008. *Fundamentals of Hydrology*. Hoboken, NJ: Taylor & Francis.
3. Ibid.
4. CECH, T. V. 2009. *Principles of Water Resources: History, Development, Management, and Policy*. Hoboken, NJ: John Wiley & Sons, Inc.
5. CLOSS, G., DOWNES, B. J. & BOULTON, A. J. 2004. *Freshwater Ecology: A Scientific Introduction*. Malden, MA: Wiley-Blackwell.
6. Ibid.
7. WARD, R. C. & ROBINSON, M. 2000. *Principles of Hydrology*. London: McGraw-Hill.
8. Ibid.
9. DAVIE, T. 2008. *Fundamentals of Hydrology*. Hoboken, NJ: Taylor & Francis.
10. WARD, R. C. & ROBINSON, M. 2000. *Principles of Hydrology*. London: McGraw-Hill.
11. Ibid.
12. Ibid.

13. Ibid.
14. Ibid.
15. Ibid.
16. Ibid.
17. DAVIE, T. 2008. *Fundamentals of Hydrology*. Hoboken, NJ: Taylor & Francis.
18. Ibid.
19. Ibid.
20. Ibid.
21. Ibid.
22. SKINNER, B. J. & PORTER, S. C. 2000. *The Dynamic Earth: An Introduction to Physical Geology*. New York: John Willey & Sons, Inc.
23. Ibid.
24. DAVIE, T. 2008. *Fundamentals of Hydrology*. Hoboken, NJ: Taylor & Francis.
25. Ibid.
26. Ibid.
27. Ibid.
28. Ibid.
29. Ibid.
30. Ibid.
31. CECH, T. V. 2009. *Principles of Water Resources: History, Development, Management, and Policy*. Hoboken, NJ: John Wiley & Sons, Inc.
32. Ibid.
33. STEPHENSON, D. 2003. *Water Resources Management*. Rotterdam: A.A. Balkema.
34. Ibid.
35. NOAA. 2011. *What is meant by the term drought?* [Online]. Available: http://www.wrh.noaa.gov/fgz/science/drought.php (accessed 2 June 2016).
36. PALANIAPPAN, M., GLEICK, P. H., ALLEN, L., COHEN, M. J., CHRISTIAN-SMITH, J. & SMITH, C. 2010. *Clearing the Waters: A Focus on Water Quality Solutions*. Nairobi: UNEP.
37. CECH, T. V. 2009. *Principles of Water Resources: History, Development, Management, and Policy*. Hoboken, NJ: John Wiley & Sons, Inc.
38. DODDS, W. K. 2002. *Freshwater Ecology: Concepts and Environmental Applications*. San Diego, CA: Academic Press.
39. DAVIE, T. 2008. *Fundamentals of Hydrology*. Hoboken, NJ: Taylor & Francis.
40. CECH, T. V. 2009. *Principles of Water Resources: History, Development, Management, and Policy*. Hoboken, NJ: John Wiley & Sons, Inc.
41. Ibid.
42. Ibid.
43. DAVIE, T. 2008. *Fundamentals of Hydrology*. Hoboken, NJ: Taylor & Francis.
44. DODDS, W. K. & WHILES, M. R. 2010. *Freshwater Ecology: Concepts and Environmental Applications of Limnology*. Cambridge, MA: Elsevier Science.
45. CHRISTCHURCH CITY COUNCIL. 2003. *Waterways, Wetlands and Drainage Guide. Part A: Visions*. Christchurch: Christchurch City Council.
46. CENTER FOR WATERSHED PROTECTION. 1997. Impact of suspended and deposited sediment. *Watershed Protection Techniques*, 2, 58–59.
47. CECH, T. V. 2009. *Principles of Water Resources: History, Development, Management, and Policy*. Hoboken, NJ: John Wiley & Sons, Inc.
48. DAVIE, T. 2008. *Fundamentals of Hydrology*. Hoboken, NJ: Taylor & Francis.
49. DODDS, W. K. 2002. *Freshwater Ecology: Concepts and Environmental Applications*. San Diego, CA: Academic Press.
50. CECH, T. V. 2009. *Principles of Water Resources: History, Development, Management, and Policy*. Hoboken, NJ: John Wiley & Sons, Inc.

51. Ibid.
52. DODDS, W. K. 2002. *Freshwater Ecology: Concepts and Environmental Applications.* San Diego, CA: Academic Press.
53. DAVIE, T. 2008. *Fundamentals of Hydrology.* Hoboken, NJ: Taylor & Francis.
54. DODDS, W. K. 2002. *Freshwater Ecology: Concepts and Environmental Applications.* San Diego, CA: Academic Press.
55. DAVIE, T. 2008. *Fundamentals of Hydrology.* Hoboken, NJ: Taylor & Francis.
56. PALANIAPPAN, M., GLEICK, P. H., ALLEN, L., COHEN, M. J., CHRISTIAN-SMITH, J. & SMITH, C. 2010. *Clearing the Waters: A Focus on Water Quality Solutions.* Nairobi: UNEP.
57. CECH, T. V. 2009. *Principles of Water Resources: History, Development, Management, and Policy.* Hoboken, NJ: John Wiley & Sons, Inc.
58. PALANIAPPAN, M., GLEICK, P. H., ALLEN, L., COHEN, M. J., CHRISTIAN-SMITH, J. & SMITH, C. 2010. *Clearing the Waters: A Focus on Water Quality Solutions.* Nairobi: UNEP.
59. BOYD, C. E. 2000. *Water Quality: An Introduction.* Boston, MA: Kluwer Academic Publishers.
60. CECH, T. V. 2009. *Principles of Water Resources: History, Development, Management, and Policy.* Hoboken, NJ: John Wiley & Sons, Inc.
61. SPELLMAN, F. R. 2008. *The Science of Water: Concepts and Applications.* Boca Raton, FL: CRC Press.
62. EPA. 1995. *Controlling nonpoint source runoff pollution from roads, highways and bridges* [Online]. Available: http://nepis.epa.gov/Exe/ZyNET.exe/P10070LL.TXT?Zy ActionD=ZyDocument&Client=EPA&Index=1995+Thru+1999&Docs=&Query=&Time=& EndTime=&SearchMethod=1&TocRestrict=n&Toc=&TocEntry=&QField=&QField Year=&QFieldMonth=&QFieldDay=&IntQFieldOp=0&ExtQFieldOp=0&XmlQuery=& (accessed 2 June 2016).
63. CHRISTCHURCH CITY COUNCIL. 2003. *Waterways, Wetlands and Drainage Guide. Part A: Visions.* Christchurch: Christchurch City Council.
64. ARNOLD, C. L. & GIBBONS, C. J. 1996. Impervious surface coverage: the emergence of a key environmental indicator. *Journal of the American Planning Association*, 62, 243–258.
65. FOSTER, S. S. D. 2001. The interdependence of groundwater and urbanisation in rapidly developing cities. *Urban Water*, 3, 185–192.
66. Ibid.
67. LERNER, D. N. 2002. Identifying and quantifying urban recharge: a review. *Hydrogeology Journal*, 10, 143–152.
68. FOSTER, S. S. D. 2001. The interdependence of groundwater and urbanisation in rapidly developing cities. *Urban Water*, 3, 185–192.
69. NATHANSON, J. A. 2008. *Basic Environmental Technology: Water Supply, Waste Management, and Pollution Control.* Upper Saddle River, NJ: Pearson Prentice Hall.
70. Ibid.
71. Ibid.
72. Ibid.
73. CECH, T. V. 2009. *Principles of Water Resources: History, Development, Management, and Policy.* Hoboken, NJ: John Wiley & Sons, Inc.
74. Ibid.
75. WHO. 2010. *WHO guidelines for drinking-water quality.* Available: http://www.who.int/water_sanitation_health/dwq/guidelines/en/ (accessed 10 May 2016).

2 What is urban water security?

Introduction

The concept of 'water security' was first introduced in the Ministerial Declarations of the Second World Water Forum in the Hague in 2000. The declarations stated water is vital for the health of humans and ecosystems and a basic requirement for the development of countries; however, water resources and related ecosystems are under threat from pollution, unsustainable use, land-use changes, climate change and other forces. As such, to achieve water security the declarations stated: water resources and related ecosystems need protecting and improving, sustainable development and political stability are to be promoted, every person needs access to enough safe water at an affordable cost and the vulnerable are protected from water-related hazards.[1]

The United Nations has defined water security as the 'capacity of a population to safeguard sustainable access to adequate quantities of acceptable quality water for sustaining livelihoods, human well-being and socio-economic development, for ensuring protection against water-borne pollution and water-related disasters, and for preserving ecosystems in a climate of peace and political stability'.[2] The key elements of achieving urban water security are in the succeeding text and summarised in Figure 2.1:

- Access to safe and sufficient drinking water at an affordable cost in order to meet basic needs including sanitation and hygiene and safeguarding of health and well-being
- Protection of livelihoods, human rights and cultural and recreational values
- Preservation and protection of ecosystems in water allocation and management systems in order to maintain their health and sustain the functioning of ecosystem services

Urban Water Security, First Edition. Robert C. Brears.
© 2017 John Wiley & Sons, Ltd. Published 2017 by John Wiley & Sons, Ltd.

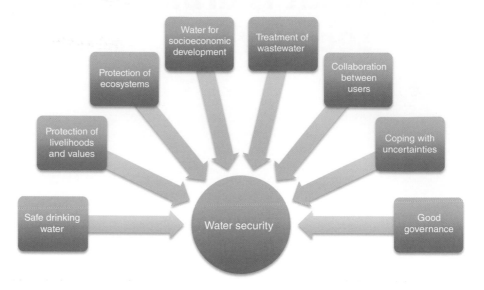

Figure 2.1 Elements of water security.

- Water supplies for socioeconomic development and activities (energy, transport, industry and tourism, etc.)
- Collection and treatment of used water to protect human life and nature from pollution
- Collaborative approaches to transboundary water resource management within and between countries to promote freshwater sustainability and cooperation
- The ability to cope with uncertainties and risk of water-related hazards, for example, floods, droughts and pollution
- Good governance and accountability and the consideration of the interests of all stakeholders through effective legal regimes; transparent, participatory and account-able institutions; properly planned, operated and maintained infrastructure; and capacity development[3]

Nonetheless, achieving water security is not a static goal; instead it is an ever-changing continuum that alters with numerous challenges, both non-climatic and climatic.[4] Therefore, future water security depends not only on meeting increased demand but also on how effectively humans can use limited water resources to meet these needs.[5]

2.1   Non-climatic challenges to achieving urban water security

Cities are at risk of water insecurity from numerous challenges including rapid population growth and urbanisation, economic growth and rising income levels and increased demand for energy and food impacting the availability of good quality water of sufficient quantity.

2.1.1 Population growth and demographic changes

The world's population is likely to grow by 30 percent between 2000 and 2025 and as much as 50 percent between 2000 and 2050. In 2011, the world's population reached seven billion and is projected to reach nine billion by 2050 with population growth occurring disproportionately in low- to middle-income countries and in urban centres: many with inadequate or barely adequate water supplies to support population levels that existed in 2000.[6]

As population growth slows and regions experience improved economic circumstances, the sizes of households are expected to shrink while the number of households increases. This will mean connections and service points for water and sanitation will need to keep up with changes in household structures.[7] Meanwhile, the age of populations will influence water consumption patterns. Increasing longevity will require greater provision of medicines and medical facilities that potentially impact the quality of water from pharmaceutical contaminants,[8] while young people, in both developing and developed countries, exposed to globalisation of trade and advertising, will be tempted to consume more products, putting increased demand on limited water supplies.[9]

2.1.2 Rapid urbanisation

Cities are important drivers of development and poverty reduction in both urban and rural areas as they concentrate most of the national economic activity, government services, commerce and transportation and provide crucial links to rural areas. Urban living is also associated with higher levels of literacy and education, better health, increased access to social services and enhanced cultural and political participation. However, rapid and unplanned urban growth threatens sustainable development when necessary infrastructure is not developed or when policies are not implemented to ensure that the benefits of city life are equitably shared. In some cities, unplanned or inadequately managed, urban expansion has led to urban sprawl, water pollution and environmental degradation together with unsustainable production and consumption patterns.[10,11]

In 2014, 54 percent of the world's population resided in urban areas. This figure is projected to increase to 66 percent by 2050. All regions around the world are expected to urbanise further, with Africa and Asia urbanising faster than all other regions, from 40 and 48 percent to 56 and 64 percent in 2050, respectively. The urban population of the world has grown rapidly from 746 million in 1950 to 3.9 billion in 2014. By 2050 the world's urban population is projected to reach 6.3 billion with almost 90 percent of that increase occurring in urban areas of Africa and Asia.[12] Meanwhile high-income countries have been highly urbanised for several decades, while upper-middle-income countries have experienced the fastest pace of urbanisation since 1950. In 1950, 57 percent of the population in high-income countries lived in urban areas. Their level of urbanisation is expected to rise from 80 percent today to 86 percent in 2050, while in 1950 only 20 percent of the population in upper-middle-income countries lived in urban areas. This has risen to 63 percent today and is projected to rise to 79 percent in 2050.[13] Meanwhile population growth is predicted for all sizes of cities: mega, large and medium.

Megacities

In 1990 there were 10 cities with populations of 10 million or more. At the time, these megacities were home to 153 million people, representing less than 7 percent of the global urban population. Today, the number of megacities has nearly tripled to 28 with a total population of 453 million, accounting for 12 percent of the world's population. By 2030, the world is projected to have 41 megacities.[14]

Large cities

Cities with populations of 5–10 million inhabitants account for a small, but growing, proportion of the global urban population. In 2014, just over 300 million people lived in 43 of these 'large' cities: 8 percent of the world's urban population. By 2030 more than 400 million people will be living in large cities, representing nearly 9 percent of the global urban population.[15]

Medium–small cities

The global population living in medium-sized cities (one to five million inhabitants) will nearly double between 2014 and 2030, from 827 million to 1.1 billion. Meanwhile, the number of people living in cities with 500 000 and one million inhabitants is expected to grow at a similar pace, increasing from 363 million in 2014 to 509 million in 2030.[16]

2.1.3   Rapid economic growth and rising income levels

By 2050, the world's economy will grow to four times its current size. While this growth will result in a less than proportional increase in water demand, the global economy will still require 55 percent more water.[17] Global water demand for manufacturing is projected to increase by 400 percent from 2000 to 2050.[18] Meanwhile the proportion of water required by industry will increase: approximately 20 percent of the world's freshwater resources are used by industry with the percentage of a country's industrial sector's water demand proportional to the average income level, ranging from around 5 percent of water withdrawals in low-income countries to over 40 percent in some high-income countries.[19] However, economic growth is dependent on the hydrological cycle with a 1 percent increase in drought area leading to a 2.8 percent reduction in economic growth, while a 1 percent increase in area impacted by floods results in a 1.8 percent decrease in economic growth.[20] Household water demand is projected to increase by 130 percent due to higher incomes and living standards.[21,22] According to HSBC, almost

3 billion people, more than 40 percent of the world's current population, will join the middle classes by 2050. As a result, emerging markets will comprise almost two-thirds of global consumption by 2050, compared to one-third today, impacting global water consumption patterns.[23] One of the main expected changes of rising incomes in emerging economies is a shift in diet from predominantly starch-based to water-intensive meat and dairy products.[24,25]

2.1.4 Increased demand for energy

The International Energy Agency (IEA) estimates that global freshwater withdrawals for energy production in 2010 were 583 billion cubic metres, some 15 percent of the world's total water withdrawals.[26] By 2040 the IEA projects global energy demand to increase by 37 percent with the world's energy supply mix divided into almost four equal parts: oil, gas, coal and low-carbon sources. Regarding fossil fuel-based energy, increased oil use for transportation and petrochemicals will see global demand for oil increase from 90 million barrels per day in 2013 to 104 million barrels per day in 2040; demand for natural gas will increase by more than half over the same time period – the fastest rate among fossil fuels – while global demand for coal will grow by 15 percent. How much water is required to meet increased demand for fossil fuel-based energy depends on whether the world follows a business-as-usual approach towards energy efficiency: following a business-as-usual approach, the IEA projects water demand for energy to be 35 percent higher than 2010, compared to a more energy-efficient future requiring 20 percent more water. Regarding low-carbon sources, global production of liquid biofuels has expanded from 16 billion litres in 2000 to more than 100 billion litres in 2011.[27] However, biofuel has significant impacts on water resources because of its water requirements during crop growth (photosynthesis) and water use in biorefineries.[28] In India, the country's biofuel programme, which focuses on *Jatropha*-based biofuel production from sugar molasses, has been constrained by water scarcity.[29] Demand for electricity is projected to increase by 70 percent between now and 2035.[30] Because thermal power generation and hydropower, which account for 80 and 15 percent of global electricity generation, are water intensive, the estimated 70 percent increase in electricity production would translate into a 20 percent increase in freshwater withdrawals.[31]

2.1.5 Increased demand for food

Agriculture accounts for 70 percent of all water withdrawn. Annual global agricultural water consumption includes crop water consumption for food, fibre and seed production plus evaporation losses from the soil and open water associated with agriculture, for example, rice fields, irrigation canals and reservoirs. By 2050, the world will require 60 percent more food produce to maintain current consumption patterns.[32] This will result in the volume of global water withdrawn for irrigation increasing from 2.6 billion cubic kilometres in 2005–2007 to 2.9 billion cubic kilometres in 2050.[33] However, the expansion of agricultural and food production to meet increased food demand from a growing global population impacts both terrestrial and aquatic ecosystems. In particular, intensive agricultural production changes the physical properties of soils reducing water infiltration rates and increasing run-off leading to lower groundwater recharge rates, soil erosion and loss of soil nutrients and increased contamination and nutrient loadings of waterways.[34] Additionally, increased demand for limited water supplies for agricultural production places pressure on water-intensive food producers to seek alternative supplies, often leading to inter-sectoral competition for limited water resources.[35] An example of agricultural production impacting both water quantity and quality is in the Middle East and North Africa region where agriculture accounts for over 85 percent of water withdrawals in many of the region's economies. Irrigation systems in the region are dependent on groundwater resources, and so declining aquifer levels and extraction of nonrenewable groundwater present an increasing risk to food production in the region. In addition, water use in agriculture is degrading water quality through the use of fertilisers and pesticides.[36]

2.2 Climatic challenges to achieving urban water security

Climate change refers to the change in state of the climate that can be statistically identified through changes in the mean and/or variability of its properties and persists for extended periods of time, usually decades or longer. Climate change is due to natural internal processes (solar cycles, volcanic eruptions) and anthropogenic activities that change the composition of the atmosphere.[37] Reasons for concern (RFCs) from climate change were first identified in the IPCC Third Assessment Report and illustrate the implications of climate change and adaptation limits for people, economies and ecosystems. There are five main RFCs:

1 *Unique and threatened systems*: The number of threatened species at risk from severe consequences of temperature rise increases significantly, particularly when temperatures rise by 2 degree Celsius above the global average (1986–2005).
2 *Extreme weather events*: Risks from climate change-related extreme weather events, including floods, droughts and heatwaves, are high with just a 1 degree Celsius increase in warming.
3 *Distribution of impacts*: Risks are evenly distributed across all countries and communities both developed and developing. Regions already experiencing decreased water availability will face increased risk with temperature increases of 2 degree Celsius.
4 *Global aggregate impacts*: There is general agreement that aggregate economic losses will accelerate as temperatures increase. If temperatures increase by 3 degree Celsius, there will likely be extensive biodiversity loss, resulting in loss of ecosystem services that not only threatens nature but also socioeconomic development.
5 *Large-scale singular events*: As temperatures increase physical systems or ecosystems may be at risk of abrupt and irreversible change, especially when temperatures rise by 1–2 degree Celsius.[38]

2.2.1   Impacts of climate change on water quality and quantity

Freshwater-related risks of climate change increase significantly with increasing greenhouse gas concentrations: the proportion of the world's population experiencing water scarcity and major flooding events will increase with rising temperatures. Climate change is projected to decrease availability of renewable surface water and groundwater resources significantly, intensifying competition for water resources among users.[39] In presently dry regions droughts will likely increase, while precipitation is projected to increase at high latitudes. With extreme weather events (floods and droughts) climate change is projected to reduce the availability of good quality water and pose threats to drinking water quality, even with conventional treatment processes, due to interacting factors including increased temperature; increased sediment, nutrient and pollutant loadings from heavy rainfall; increased concentrations of pollutants during droughts; and disruption of treatment facilities during floods.[40] With climate change, water utilities will be confronted with the following impacts:

- *Increased temperature*: Decreasing snow/ice volumes and increasing evaporation rates from lakes, reservoirs and aquifers will decrease the quantity of water available to users. In addition, increased temperatures will increase demand for water.
- *Shifts in timing of river flows*: More frequent and intense droughts will increase the need for artificial water storage.
- *Higher water temperatures*: Increased algal blooms and natural organic material will lead to water needing additional or new treatment processes for drinking water.
- *Drier conditions*: Increasing pollutant concentrations will threaten groundwater supplies that are already of low quality (high concentrations of arsenic, iron and manganese).
- *Increased storm water run-off*: Increased storm water run-off will increase loads of pathogens, nutrients and suspended sediment.
- *Sea-level rise*: Increased sea-level rises will increase the salinity of coastal aquifers particularly when groundwater recharge is predicted to decrease.[41]

In addition, wastewater treatment technologies vary in their resilience to climate change impacts. With sewage systems there are three climatic conditions of interest:

- *Wet weather*: During periods of heavy precipitation, there are increased amounts of storm water and wastewater entering combined systems for short periods. As such, current designs, based on 'design storms' that are defined through analysis of historical precipitation data, need to be redesigned to increase their capacity.
- *Dry weather*: During periods of dry weather and droughts, soil shrinks as it dries, causing water mains and sewage pipes to crack making them vulnerable to infiltration and exfiltration of water and wastewater. The combined effects of higher temperatures, increased pollutant concentrations, longer retention times and sedimentation of solids may lead to corrosion of sewers, shorter asset lifespans, more drinking water pollution and higher maintenance costs.
- *Sea-level rise*: Intrusion of salt water into the sewers will necessitate processes that can handle saltier wastewater.[42]

2.2.2   Socioeconomic risks of climate change

According to the IPCC, climate change extreme weather events and temperature rises, which affect the quality and quantity of water resources, will expose individuals, communities and countries to numerous socioeconomic risks:

- *Global economy*: Economic forecasts on the impacts of climate change vary depending on which economic subsectors are covered and whether they account for catastrophic changes and tipping points. Incomplete estimates of annual global economic losses for additional temperature increases of around 2 degree Celsius are between 0.2 and 2 percent of global income. However, these economic losses are more likely to be greater than smaller, with large variations in losses between, and within, countries,[43] for example, a study on the impacts of climate change and economic growth in Egypt found that in the absence of climate change adaptation, real GDP in 2050 will be 6.5 percent lower than without climate change.[44]
- *Human health*: Until the mid- twenty-first century, climate change is projected to impact human health mainly by exacerbating existing health problems. Examples include injury, disease and death due to more intense heatwaves, undernutrition from diminished food production especially in poor regions and increased risk from waterborne diseases.[45]
- *Human security*: Climate change is projected to increase the number of displaced people. The probability of displacement increases when populations lacking resources for planned migration experience higher exposures to extreme weather events. Climate change can increase the likelihood of conflicts by amplifying existing difficulties including economic weakness, lack of adequate infrastructure and weak governance.[46] Meanwhile, the impacts of climate change on critical infrastructure and territorial integrity are expected to influence national security policies particularly when resources are shared across boundaries.[47]
- *Livelihoods and poverty*: Climate change is projected to slow down economic growth making poverty reduction more difficult by prolonging existing, and creating new, poverty traps, particularly in developed and developing countries with rising economic inequality.[48]

2.3   Reducing non-climatic and climatic risks to urban water security

Non-climatic and climatic challenges pose four specific water risks to achieving urban water security: risk of shortage (lack of sufficient water to meet demand for water by all users), risk of inadequate quality (lack of good quality water for various uses), risk of excess (overflow of the water system and damage to infrastructure) and risk of undermining the resilience of natural water systems (exceeding the coping ability of natural ecosystems from excessive water withdrawals that cause irreversible damage to groundwater and surface water supplies).[49]

Water managers can use a risk management framework to create or enhance resiliency of the water system to uncertainty from non-climatic and climatic

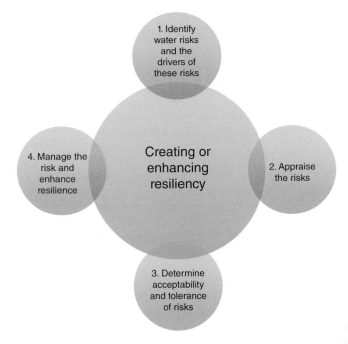

Figure 2.2 Risk management framework to create or enhance resiliency of the water system.

changes to achieve urban water security, where a resilient water system withstands service failure as much as possible and recovers from it if and when it occurs.[50,51,52] The framework is composed of four main components and is summarised in Figure 2.2:

1 *Identifying water risks and the drivers of these risks*: Risks to water resources occur from socioeconomic and socio-demographic changes and climatic change. Individuals, communities and societies all view risks differently. In water resources each user of water has a different view as to what is a risk: residential areas may see flooding as a risk, while an agricultural producer may see annual flooding as necessary for increasing food production. Alternatively people may welcome a drier, warmer summer, while energy providers may prefer wetter, cooler conditions for electricity generation.

2 *Appraising the risks*: Building up an information base to inform decisions about water risks requires bringing together a scientific risk assessment and the understanding of risk perceptions by stakeholders. While scientific and technical inputs are important elements in appraising risk, economic, social and cultural dimensions need to be considered to make sure policy responses for ensuring water security are proportional, economically efficient and socially equitable.

3 *Determining acceptability and tolerance of risks*: Determining acceptability and tolerance of risks to water security is challenging as well as controversial as it relies on both an evidence- and values-based approach. A risk is considered acceptable if the likelihood of exceeding a specific risk threshold is low and the impact of exceeding that threshold is low. Therefore there is no pressure to reduce

acceptable risks further unless cost-effective measures can be found. However, cost-effective measures are required to reduce tolerable risks to an acceptable standard.

4 *Managing the risk and enhancing resilience*: The strategy to minimise risks to water security should be informed by all the previous steps of the risk management framework. The strategy to manage risk could be to avoid, reduce or bear the risk. This can be achieved, for example, by limiting the population's exposure to the risk or enhancing the resiliency of the community, physical assets and environment by making them less vulnerable.[53]

Notes

1. WORLD WATER COUNCIL. 2000. *Ministerial declaration of The Hague on water security in the 21st century* [Online]. Available: http://www.worldwatercouncil.org/fileadmin/world_water_council/documents/world_water_forum_2/The_Hague_Declaration.pdf (accessed 9 May 2016).
2. UN-WATER. 2013. *Water security and the global water agenda*. Available: http://www.unwater.org/downloads/watersecurity_analyticalbrief.pdf (accessed 9 May 2016).
3. Ibid.
4. GWP/OECD. 2015. *Securing water sustaining growth* [Online]. Available: http://www.gwp.org/Global/About%20GWP/Publications/The%20Global%20Dialogue/SECURING%20WATER%20SUSTAINING%20GROWTH.PDF (accessed 9 May 2016).
5. UNESCO 2012. *Managing Water under Uncertainty and Risk*. Paris: UNESCO.
6. SCHUSTER-WALLACE, C.J. & SANDFORD, R. 2015. *Water in the world we want* [Online]. United Nations University Institute for Water, Environment and Health and United Nations Office for Sustainable Development. Available: http://inweh.unu.edu/wp-content/uploads/2015/02/Water-in-the-World-We-Want.pdf (accessed 9 May 2016).
7. Ibid.
8. SHENG, C., NNANNA, A. G. A., LIU, Y. & VARGO, J. D. 2016. Removal of trace pharmaceuticals from water using coagulation and powdered activated carbon as pretreatment to ultrafiltration membrane system. *Science of the Total Environment*, 550, 1075–1083.
9. UNESCO. 2009. *The United Nations World Water Development Report 3: water in a changing world* [Online]. Available: http://unesdoc.unesco.org/images/0018/001819/181993e.pdf (accessed 9 May 2016).
10. UNITED NATIONS DEPARTMENT OF ECONOMIC AND SOCIAL AFFAIRS POPULATION DIVISION. 2014. *World Urbanization Prospects: the 2014 revision, highlights* [Online]. Available: http://esa.un.org/unpd/wup/highlights/wup2014-highlights.pdf (accessed 9 May 2016).
11. ZHANG, X. Q. 2016. The trends, promises and challenges of urbanisation in the world. *Habitat International*, 54(Part 3), 241–252.
12. UNITED NATIONS DEPARTMENT OF ECONOMIC AND SOCIAL AFFAIRS POPULATION DIVISION. 2014. *World Urbanization Prospects: the 2014 revision, highlights* [Online]. Available: http://esa.un.org/unpd/wup/highlights/wup2014-highlights.pdf (accessed 9 May 2016).
13. Ibid.
14. Ibid.
15. Ibid.
16. Ibid.

17. UN-WATER. 2014a. *Partnerships for improving water and energy access, efficiency and sustainability* [Online]. Available: http://www.un.org/waterforlifedecade/water_and_energy_2014/pdf/water_and_energy_2014_final_report.pdf (accessed 9 May 2016).

18. OECD. 2012. *OECD Environmental Outlook to 2050: The Consequences of Inaction.* New York: OECD.

19. UNESCO 2012. *Managing Water under Uncertainty and Risk.* Paris: UNESCO.

20. UN-WATER. 2014b. *A post-2015 global goal for water: recommendations from un-water* [Online]. Available: http://www.un.org/waterforlifedecade/pdf/27_01_2014_un-water_paper_on_a_post2015_global_goal_for_water.pdf (accessed 9 May 2016).

21. HARLAN, S. L., YABIKU, S. T., LARSEN, L. & BRAZEL, A. J. 2009. Household water consumption in an Arid City: affluence, affordance, and attitudes. *Society & Natural Resources*, 22, 691–709.

22. UN-WATER. 2014a. *Partnerships for improving water and energy access, efficiency and sustainability* [Online]. Available: http://www.un.org/waterforlifedecade/water_and_energy_2014/pdf/water_and_energy_2014_final_report.pdf (accessed 9 May 2016).

23. HSBC. *Consumer in 2050. The rise of the EM middle class.* Available: https://www.hsbc.com.vn/1/PA_ES_Content_Mgmt/content/vietnam/abouthsbc/newsroom/attached_files/HSBC_report_Consumer_in_2050_EN.pdf (accessed 9 May 2016).

24. KEARNEY, J. 2010. Food consumption trends and drivers. *Philosophical Transactions of the Royal Society of London B: Biological Sciences*, 365, 2793–2807.

25. UNESCO. 2012. *Managing Water under Uncertainty and Risk.* Paris: UNESCO.

26. IEA. 2012. *Water for energy. Is energy becoming a thirstier resource?* [Online]. Available: http://www.worldenergyoutlook.org/media/weowebsite/2012/WEO_2012_Water_Excerpt.pdf (accessed 9 May 2016).

27. IRENA. 2015. *Renewable energy in the water, energy and food nexus* [Online]. Available: http://www.irena.org/documentdownloads/publications/irena_water_energy_food_nexus_2015.pdf (accessed 9 May 2016).

28. UNESCO. 2012. *Managing Water under Uncertainty and Risk.* Paris: UNESCO.

29. MIRZABAEV, A., GUTA, D., GOEDECKE, J., GAUR, V., BÖRNER, J., VIRCHOW, D., DENICH, M. & VON BRAUN, J. 2015. Bioenergy, food security and poverty reduction: trade-offs and synergies along the water–energy–food security nexus. *Water International*, 40, 772–790.

30. UNESCO. 2015. *United Nations World Water Development Report 2015: water for a sustainable world* [Online]. Available: http://unesdoc.unesco.org/images/0023/002318/231823E.pdf (accessed 9 May 2016).

31. Ibid.

32. UNESCO. 2012. *Managing Water under Uncertainty and Risk.* Paris: UNESCO.

33. FAO AND WWC. 2015. *Towards a water and food secure future. Critical perspectives for policy-makers* [Online]. Available: http://www.fao.org/3/a-i4560e.pdf (accessed 9 May 2016).

34. SCHUSTER-WALLACE, C.J. & SANDFORD, R. 2015. *Water in the world we want* [Online]. United Nations University Institute for Water, Environment and Health and United Nations Office for Sustainable Development. Available: http://inweh.unu.edu/wp-content/uploads/2015/02/Water-in-the-World-We-Want.pdf (accessed 9 May 2016).

35. BHADURI, A., RINGLER, C., DOMBROWSKI, I., MOHTAR, R. & SCHEUMANN, W. 2015. Sustainability in the water–energy–food nexus. *Water International*, 40, 723–732.

36. ANTONELLI, M. & TAMEA, S. 2015. Food-water security and virtual water trade in the Middle East and North Africa. *International Journal of Water Resources Development*, 31, 326–342.

37. IPCC. 2014a. *AR5 summary for policymakers.* Available: http://ipcc-wg2.gov/AR5/report/ (accessed 9 May 2016).

38. Ibid.
39. Ibid.
40. Ibid.
41. IPCC. 2014b. *Mitigation of climate change.* Available: http://www.ipcc-wg2.gov/ (accessed 9 May 2016).
42. Ibid.
43. IPCC. 2014a. *AR5 summary for policymakers.* Available: http://ipcc-wg2.gov/AR5/report/ (accessed 9 May 2016).
44. ELSHENNAWY, A., ROBINSON, S. & WILLENBOCKEL, D. 2016. Climate change and economic growth: an intertemporal general equilibrium analysis for Egypt. *Economic Modelling,* 52, Part B, 681–689.
45. IPCC. 2014a. *AR5 summary for policymakers.* Available: http://ipcc-wg2.gov/AR5/report/ (accessed 9 May 2016).
46. HERMAN, P. F. & TREVERTON, G. F. 2009. The political consequences of climate change. *Survival,* 51, 137–148.
47. IPCC. 2014a. *AR5 summary for policymakers.* Available: http://ipcc-wg2.gov/AR5/report/ (accessed 9 May 2016).
48. Ibid.
49. OECD. 2013b. *Water security for better lives: a summary for policymakers.* Available: http://www.oecd.org/env/resources/Water%20Security%20for%20Better%20Lives-%20brochure.pdf (accessed 9 May 2016).
50. BUTLER, D., FARMANI, R., FU, G., WARD, S., DIAO, K. & ASTARAIE-IMANI, M. 2014. A new approach to urban water management: safe and sure. *Procedia Engineering,* 89, 347–354.
51. IPCC. 2014a. *AR5 summary for policymakers.* Available: http://ipcc-wg2.gov/AR5/report/ (accessed 9 May 2016).
52. OECD. 2013a. *Water and climate change adaptation policies to navigate uncharted waters.* Available: http://www.oecd.org/publications/water-and-climate-change-adaptation-9789264200449-en.htm (accessed 2 June 2016).
53. Ibid.

3 Managing water sustainably to achieve urban water security

Introduction

Transitions towards urban water security not only consist of physical and technological changes but also involve behavioural change, the purpose being to transition towards a society that manages water resources sustainability[1] – in particular, a society that balances the demand for water resources with scarce supplies. However, the question is: what does sustainability mean? While the term 'sustainability' has become a buzzword in various multilateral reports, media and political commentary, there is in fact neither unanimous international definition of what the term means and how it can be achieved, nor is there any definition as to what sustainability looks like physically.

This chapter first defines what sustainability and sustainable development are before discussing the differing approaches to sustainability. The chapter then introduces sustainable water management frameworks to achieve water security: integrated water resources management (IWRM) at the river basin level and integrated urban water management (IUWM) at the urban level. Finally, the chapter discusses how IUWM can achieve urban water security.

3.1 What is sustainability?

The Brundtland Report states that sustainability is a 'development that fulfils the needs of the present generation, without compromising the ability of future generations to fulfil their needs'.[2] In particular, the report attempts to

Urban Water Security, First Edition. Robert C. Brears.
© 2017 John Wiley & Sons, Ltd. Published 2017 by John Wiley & Sons, Ltd.

Table 3.1 Differing values placed on ecosystems

Ecosystem value	Description
Direct use value	The extractive, consumptive or structural use value of ecosystem goods or services that can be extracted, consumed or enjoyed directly, for example, drinking water
Indirect value	Services ecosystems provide, for example, water purification
Option value	Value attached to maintaining the possibility of obtaining benefits from ecosystem goods and services in the future
Non-use value	The existence value – the value people derive from the knowledge that something exists even if they never plan on seeing or using it
Bequest value	The value derived from the desire to pass on to future generations viable, healthy ecosystems

connect environmental, economic and social aspects of sustainability into the concept of sustainable development, with the objective of sustainable development being the maximisation of each pillar in development.[3,4]

However, while the two terms 'sustainability' and 'sustainable development' have become popular in policy-oriented research, as what policies ought to achieve, the real question is: what exactly do these terms mean? In fact, the terms have come to mean different things to different people. As such, 'sustainability' and 'sustainable development' are laden with actual or potential conflicts, running contrary to the majority of discourse that assumes these concepts will generate desirable outcomes for all, all the time.[5,6,7,8] In particular, there are numerous conflicts between individuals in society on what to sustain as each person values the environment differently. Table 3.1 lists the numerous values people place on the ecosystem.[9]

3.1.1   Urban sustainability

The concept of sustainability, post-Brundtland, has spread throughout a variety of policy areas including urban policy. Globally, national and local governments are implementing policies to make cities more sustainable. The major challenge of urban sustainability is to ensure economic, social and environmental sustainability now and into the medium–long-term future.[10] Urban sustainability, therefore, is about creating a balanced relationship between urban and rural areas, conservation of natural spaces in urban areas and environmental management to reduce pollution and promote conservation.[11]

Urban centres have physical footprints in that they occupy around 2 percent of the Earth's surface. However, with globalisation creating a global exchange of goods and services, cities have become the main consumers of the planet's ecosystems goods and services: cities consume 75 percent of the world's resources including water, food, energy, forestry and construction materials and produce more than 50 percent of global waste.[12,13,14] As such, urban centres have an ecological footprint, which is the geographical area required to produce the quantity of any resource or ecological service used by a defined population or economy.[15] This has led to urban centres becoming reliant on 'imports' of natural resources from other ecosystems, most of which are often located at great distances from urban centres.[16] However, this creates dependencies with other regions that may not be ecologically, economically or geopolitically viable in the future.[17,18]

To achieve sustainability, where a true sustainable city is one where its ecological footprint matches more or less its physical footprint, cities need to ensure that resources imported from distant geographic regions are sustainably used by minimising resource consumption and that pollution does not exceed nature's regeneration capacity.[19,20,21,22] This recognises the ecological limits of a finite plant.[23] To achieve the goal of sustainability, urban centres can utilise their concentration of infrastructure and services (economies of scale) to increase the efficient use of resources, promote the recycling and reuse of resources and decouple population growth from resources use.[24]

3.1.2 Approaches to sustainability

Sustainability proponents can be divided into those that adhere to weak sustainability and those adhering to strong sustainability.[25] In the weak form of sustainability, natural capital is substitutable for other types of capital in the pursuit of economic growth, while strong sustainability places a priority over the maintenance or improvement of current levels of natural capital in the pursuit of economic growth.[26]

In weak sustainability there is no difference between natural and other forms of capital. As long as the natural capital being depleted is replaced by even more valuable physical and human capital, then the aggregate stock (human, physical and remaining natural capital) is, at the minimum, being retained for future generations.[27,28,29] Weak sustainability is based around a key assumption that new technologies, fostered through appropriate market instruments, can reduce environmental degradation. As such, proponents of strong sustainability argue that weak sustainability ignores the human, social and cultural drivers of environmental degradation.[30] Furthermore, opponents of weak sustainability argue that it promotes a 'take–make–waste' economic framework where natural resources are taken from the environment, converted into goods and services from which large amounts of waste is returned back into the environment causing irreversible environmental damage.[31] This economic model has led to rapid accumulation of physical and human capital and excessive depletion and degradation of natural capital.[32,33]

Strong sustainability proponents argue that natural capital is not substitutable with other forms of capital for three reasons: first, the depreciation of natural capital is irreversible or takes long periods of time to recover; second, it is not possible to replace a depleted resource with a new one; and third, ecosystems can collapse abruptly.[34,35] As such, supporters of strong sustainability believe that natural capital cannot be substituted with other forms of capital; in particular, no amount of physical or human capital can replace all the environmental resources that comprise natural capital or the ecological services performed by nature. Therefore, natural capital should be protected, not depleted.[36,37]

Table 3.2 Ecosystem services

Ecological services	Description	Examples
Provisioning services	Products obtained from ecosystems	Food, water, fuelwood, fibre, biochemicals, genetic resources
Regulating services	Benefits obtained from the regulation of ecosystem processes	Climate regulation (maintenance of temperatures), water regulation (flood protection), water purification
Supporting services	Necessary for the continuation of the three other types of ecosystem services (provisioning, regulating, cultural)	Soil formation, nutrient cycling, primary production
Cultural services	Non-material benefits obtained from ecosystems	Spiritual, religious, recreational, aesthetic, inspirational, educational, cultural

VOORA, V. A. AND VENEMA, H. D. 2008. *The Natural Capital Approach: A Concept Paper.* International Institute for Sustainable Development, Winnipeg, Manitoba, Canada

3.1.3   Environmental pillar of strong sustainability

In strong sustainability, the environmental pillar seeks to protect the integrity of natural ecosystems and the various ecosystem services necessary for human survival.[38,39] In particular, strong sustainability recognises that underlying all the resources humans consume are ecosystem services, which are defined as the benefits that people obtain from ecosystems such as clean air, purified water, etc.[40,41] Specifically, there are four types of services that ecosystems provide: provisioning, regulating, supporting and cultural services (Table 3.2). Therefore, to achieve environmental sustainability in the strong form, society needs to value the various services ecosystems provide by enabling ecosystems to regenerate by reducing current human levels of exploitation, relieving pressure on ecosystems by investing in projects that minimise that pressure and improving efficiency in the use of ecosystems.[42]

3.1.4   Economic pillar of strong sustainability

In the economic pillar of strong sustainability, natural capital is valued as it provides the vast majority of goods and services humans rely on for survival and economic growth, for example, food, fuel, construction materials and purification of air and water.[43] In recognising this value, it enables individuals and society as a whole to frame consumption choices and make clear trade-offs between various outcomes.[44,45] Therefore, in strong sustainability, natural capital is recognised as a diminishing resource, and as economics concerns the efficient use of scarce resources, there is a role for economics in valuing biodiversity.[46]

Nonetheless, there are objections to valuing nature: first, it is a strictly anthropogenic measure which does not account for non-human values and needs and, second, pricing of the natural world is seen as an example of the moral failings of the capitalist system where everything is thought of as a commodity with monetary value.[47] However, Boyer and Polasky[48] reject the first notion and argue that the alternative is the ecocentric view where the source of value may be other species or ecosystem processes rather than how species or ecosystems satisfy human needs and wants. Regarding the second objection, the authors argue the point of valuing nature is not to think in monetary terms but to frame choices, enabling society to make clear trade-offs between various outcomes.[49]

Overall, the strong sustainability economic model is one based on a borrow–use–replenish framework where resources are converted into energy, goods and services with the by-product either returned to the economy for future use or returned back to nature as nutrients for further use.[50] This model ensures unnecessary waste is avoided and pollution does not exceed the regenerative capacity of nature.[51]

3.1.5 Social pillar of strong sustainability

The social pillar of strong sustainability recognises that an unjust society is unlikely to be sustainable in environmental or economic terms. Rather, social tensions are likely to undermine the recognition by citizens of both their environmental rights and duties relating to environmental degradation.[52] Therefore, a better understanding of sustainable development's concept of social sustainability is critical for reconciling the competing demands of the society–environment–economic tripartite.[53]

In the context of environmental sustainability, there are five interconnected equity principles of social sustainability:

1 *Intergeneration equity*: This is equity between generations where the future generation's standards of living should not be disadvantaged by the activities of the current generation's standard of living.
2 *Intragenerational equity*: This is equity among the current generation and can be achieved through widespread political participation by citizens.
3 *Geographical equity* (transfrontier responsibility): Whereby local policies should be geared towards resolving local and global environmental problems as political/administrative boundaries are frequently used to shield polluters from prosecution in other jurisdictions.
4 *Procedural equity*: Regulatory systems should be devised to ensure transparency as it is critical that people have the right to access environmental information on activities that have both local and global impacts.
5 *Interspecies equity*: This notion places the survival of other species on an equal basis to the survival of humans. This is to reflect the critical importance of preserving ecosystems and maintaining biodiversity for human survival. Specifically, humans have an obligation to ensure ecosystems are not degraded beyond their regenerative capacity.[54,55,56,57,58]

3.1.6   Urban resilience and sustainability

In the twenty-first century, cities need to be resilient to be sustainable[59] where urban resilience refers to the ability of a city to withstand a wide array of shocks and stresses from climate change as well as environmental degradation. According to the Asian Development Bank, resilient cities that respond to climatic and non-climatic stresses and shocks demonstrate a series of qualities through their systems and numerous stakeholders that include:

- *Reflective*: People and institutions systematically learn from experiences. They have mechanisms to continuously modify actions based on emerging evidence rather than seeking permanent solutions based on an assessment of today's shock and stresses.
- *Robust*: Robust cities are designed and managed to withstand the impacts of extreme conditions and avoid catastrophic collapses from the failure of a single element. A robust system anticipates system failure and makes provisions to maximise predictability and safety.
- *Redundant*: Redundancy is to deliberately plan capacity to accommodate for increasing demand or extreme pressures. If one component of the system fails, other substitutable components can meet essential needs.
- *Flexible*: A flexible system can change, evolve and adopt alternative strategies in the short or long term in response to changing conditions. These systems tend to favour decentralisation of conventional infrastructure with new technologies.
- *Resourceful*: People and organisations should invest in capacities to anticipate future conditions, set priorities and mobilise resources including human, financial and physical. Resourceful cities can respond quickly to extreme events by modifying organisations or procedures as required.
- *Inclusive*: Inclusion involves the consultation and engagement of the community, particularly the vulnerable members of society. This ensures resilience has collective ownership with a joint vision shared by all.
- *Integrated*: City systems, decision-making and investments should be mutually supportive of a common outcome with an ongoing feedback of information to guide further integration.[60]

3.2   What does sustainability mean in urban water management?

Cities are not only significant consumers of raw water in their own river basins but frequently of basins further away.[61] Specifically, cities directly impact the water quantity, quality and hydrological cycle of ground and surface water located both within the city limits and in surrounding areas, many of which are trans-boundary (crossing municipal, regional, national or even international boundaries). This impact is known as a 'water footprint'.[62,63] As such, cities have a strategic interest, environmentally, economically, socially and even politically, in maintaining the health and vitality of river basins they rely on.[64]

Whether water consumption is sustainable or not depends on the time it takes for the resource to regenerate compared with its usage. Water's regeneration time is rc:rs, in which rc is the rate of consumption and rs the rate of supply. Freshwater resources can be considered renewable if carefully managed; however, they can be considered non-renewable and unsustainable if they are overexploited or 'mined' (rc/rs > 1). Meanwhile, water use can be considered sustainable if the resource is utilised at a rate in which supply is greater than the amount consumed (rc/rs < 1). With increases in climate change-induced drought and scarcity, along with increases in population, it is common for consumption to be greater than supply (rc > rs), and therefore unsustainable, in many parts of the world.[65]

The most appropriate form of sustainability in urban water management is the strong sustainability viewpoint for three reasons: first, strong sustainability ensures current and future generations can meet their basic water needs; second, strong sustainability ensures there is sufficient water to produce goods and services; and third, strong sustainability ensures there is adequate quality and quantity of water resources necessary to protect ecosystems.[66] Therefore, strong sustainability reduces the potential for conflicts and tensions between the environmental, economic and social pillars of sustainable development.[67,68,69,70]

3.2.1 Environmental pillar in strong sustainable urban water management

In strong sustainability the environmental pillar of sustainable urban water management aims to protect the quality and quantity of water necessary for the survival of both humans and nature.[71,72] In particular, the environmental pillar recognises the need to protect the numerous services provided by ecosystems that are beneficial to humans and nature:

- *Provisioning services*: Services focused on directly supplying food and non-food products from water flows (freshwater supplies, crop production, hydropower, timber, livestock, etc.)
- *Regulatory services*: Services related to regulating flows or reducing hazards related to water flows – regulation of hydrological flows (buffer runoff, soil water infiltration, groundwater recharge), natural hazard mitigation (flood prevention, landslide prevention, etc.), soil protection and control of ground and surface water quality
- *Supporting services*: Services provided to support habitats and ecosystem functioning (wildlife habitat, flow regime required to maintain downstream habitat and uses)
- *Cultural and amenity services*: Services related to recreation and human inspiration (aquatic recreation, landscape aesthetics, cultural heritage and identity, artistic and spiritual inspiration)[73]

Table 3.3 Ecosystem service value

Ecosystem service	Service provided	Monetary value (US$ per hectare per year)
Provisioning service	Water for people	45–7500
Regulating service	Water quality control	60–6700
Cultural and amenity services	Recreation/tourism	230–3000

SMITH, M., DE GROOT, D. & BERGKAMP, G. 2006. *Pay: Establishing Payments for Watershed Services.* Gland: IUCN

3.2.2 Economic pillar in strong sustainable urban water management

In the strong form of sustainability, water is allocated in the most efficient way with a priority placed on uses that provide the highest value to society as a whole.[74,75] In particular, water is a special economic good with no substitute, and therefore its allocation is a societal question not a market question. As such, water is not priced solely through market forces; instead, the price of water should, first, include the full economic cost of providing the water service and, second, provide a clear signal to users that water is a scarce good that provides valuable ecosystem services (examples in Table 3.3) and should be conserved.[76]

3.2.3 Social pillar in strong sustainable urban water management

In the social pillar of strong sustainability, water is managed in a way that ensures both current and future generations have access to good quality water of sufficient quantity. In particular, the social pillar ensures there are both intergenerational and intragenerational equities as well as geographical, procedural and interspecies equities in water supplies:

- *Intergenerational equity (equity of current and future generations)*: The sustainable use of water ensures the satisfying of needs for both current and future generations. To reduce intergenerational competition over water resources, the use of water resources should not exceed the limits of its natural recharge rate so future use is safeguarded.
- *Intragenerational equity (equity among the current generation)*: Each water user has a basic right to water of adequate quantity and quality. Water users should avoid unnecessary use through the promotion of water conservation to avoid welfare losses for both current and future generations.
- *Geographical equity (transfrontier responsibility)*: River basins are often trans-boundary with water flowing over administrative and political boundaries, and so there is a responsibility to ensure all users and uses are treated equitably.

- *Procedural equity (right to environmental information)*: Regulatory systems should be devised to ensure transparency as it is critical that people have the right to access information on water quality and quantity.
- *Interspecies equity (equity between all species)*: Humans have an obligation to ensure there are adequate quantities of water of good quality sufficient for the survival of ecosystems.[77,78,79]

3.3   Sustainable water resources management frameworks

Natural resources management frameworks provide a theoretical foundation of how knowledge is generated for the purpose of effectively managing a natural resource.[80] For a management framework to be of value, first, it needs to be based on correct causal understandings of the natural resource phenomenon concerned; in particular, the understanding is based on reliable scientific theory that is translated into processes for producing and applying knowledge about management interventions of that phenomenon. Second, it needs to be 'testable' in that managers can empirically test the relationships between knowledge production and natural resource outcomes.[81]

3.3.1   Integrated water resources management

The management framework for water resources management is known as IWRM, which is a cross-sectoral approach designed to promote the coordinated development and management of water, land and related resources in order to maximise economic and social welfare in an equitable manner, without compromising the sustainability of ecosystems and the environment.[82]

IWRM is based on the understanding that water resources are an integral component of the ecosystem, a natural resource and a social and economic good. For IWRM to be successful, it requires the coordinated development and management of land and water use, surface water and groundwater, water quantity and quality, upstream and downstream use and freshwater and coastal waters, while recognising all users are interdependent on one another.[83] There are numerous examples of how users of water can affect one another, for example, high irrigation demands and polluted waterways from agriculture mean less freshwater for drinking and industrial use, while contaminated municipal and industrial wastewater pollutes rivers and threatens ecosystems.

An important aspect of IWRM is the participation of individuals and communities in all aspects of water management policy and decision-making. This ensures all members of society benefit from the sustainable and equitable use of water resources.[84] IWRM is also about modifying human systems to encourage people to use water resources sustainably.[85] There are five key principles of IWRM:

1 Freshwater is a finite and vulnerable resource, and it is essential to sustain life, development and the environment.

2 Water development and management should be based on a participatory approach involving users, planners and policymakers at all levels.
3 Women play a central role in the provision, management and safeguarding of water.
4 Water is a public good and has a social and economic value in all its competing uses.
5 IWRM is based on equitable and efficient management and sustainable use of water.[86]

3.3.2 Origins of IWRM principles

The origin of the five IWRM principles is based on the Dublin Principles developed at the *International Conference on Water and the Environment* in Dublin, Ireland, held on 26–31 January 1992. The conference participants, comprising 500 participants from 100 countries and representatives of 80 international, intergovernmental and non-governmental organisations, recognised an emerging global water crisis and called for a fundamental new approach to the assessment, development and management of freshwater resources.[87] To manage water sustainably, the participants called for governments to invest in water infrastructure, conduct public awareness campaigns, implement suitable legislative and institutional changes, invest in technology development and initiate capacity building programmes.[88]

The Dublin Statement set out four guiding principles for managing freshwater resources:

1 *Freshwater is a finite and vulnerable resources, essential to sustain life, development and the environment*: Because water sustains life, effective management of water resources requires an approach that links social and economic development with protection of natural ecosystems.
2 *Water development and management should be based on a participatory approach, involving users, planners and policymakers at all levels*: The participatory approach ensures there is awareness of the importance of water among policymakers and the general public. It means that decisions involve full public consultation and all users participate in the planning and implementation of water projects.
3 *Women play a central part in the provision, management and safeguarding of water*: Despite women playing a key role in the collection and safeguarding of water for domestic purposes and agricultural use, women in most countries play minimal roles in the development and management of water resources. Therefore, policies are required to address women's needs and empower women to participate in the decision-making and implementation process of water resources management.
4 *Water has an economic value in all its competing uses and should be recognised as an economic good*: It is vital to recognise the basic right of all humans to have access to clean water and sanitation at an affordable price. However, by managing water as an economic good, it is an important way of achieving efficient and equitable use of the resource while encouraging conservation and protection of water resources.

3.3.3 Benefits of managing water in an integrated manner

According to the Dublin Statement, there are significant benefits from managing water in an integrated way. For developing countries, in particular, effective water management will alleviate poverty and disease through increased food production, clean water and sanitation. In both developing and developed countries, there will be greater protection against natural disasters as higher quality data for water resources management enables countries to better manage droughts and floods. Regarding water conservation and sustainable urban development, effective water resources management will lead to increased water conservation and reductions in discharges of municipal and industrial wastes. This can be achieved through the enactment of appropriate water charges and discharge guidelines. In rural communities, the application of water-saving technology and improved management methods will result in increased agricultural production and rural water supply. At the same time, a more integrated approach to water resources management will protect fragile aquatic ecosystems through decreased disruption of waterways and lower levels of pollution.[89]

Raising awareness of water issues among communities is an essential aspect of IWRM as it creates a more water-aware society and increases the knowledge base for managing water.[90] For example, practical education programmes such as measuring water quantity and quality and determining environmental factors affecting water provide youth with the ability to understand impacts on water resources and users. A more integrated approach to water resources management also promotes the capacity building of local organisations enabling personnel to undertake water resources assessment and management projects, increasing the effectiveness of institutions to plan and implement water projects.[91]

3.3.4   Agenda 21 and IWRM

The Dublin Principles were presented to world leaders at the *United Nations Conference on Environment and Development* held in Rio de Janeiro, Brazil, in 1992. The outcome of this conference was Agenda 21, a comprehensive plan of action to be taken globally, nationally and locally by organisations including the United Nations, national governments and local organisations in every area in which humans impact the environment.

The importance of water resources management was recognised in Agenda 21 with seven programmes proposed for the freshwater sector: Integrated water resources development and management; water resources assessment; protection of water resources, water quality and aquatic ecosystems; drinking water supply and sanitation; water and sustainable urban development; water for sustainable food production and rural development and impacts of climate change on water resources.[92]

Concerning IWRM, Agenda 21 outlines four objectives that should be pursued by all countries: objective one promotes the need for an interactive and multisectoral approach to water resources management along with the identification and protection of sources of freshwater supply that integrates technological, socioeconomic, environment and human health considerations. Objective two states the need to plan for the sustainable and rational utilisation, protection, conservation and management of water resources based on community needs and priorities. Objective three promotes the need for full public participation including women, youth, indigenous people and local communities in water management policy and decision-making. Objective four recommends countries to strengthen appropriate institutional, legal and financial mechanisms to ensure water policy promotes sustainable social progress and economic growth.[93]

3.3.5   The role of efficiency in IWRM

In 2002, the World Summit on Sustainable Development (WSSD) recognised the need of improving water efficiency. Article 26 of the WSSD Plan of Implementation makes reference to water efficiency in two different ways. In part A, countries should introduce measures to improve the efficiency of water infrastructure in order to reduce losses and increase the recycling of water, while part B states the need to promote allocation efficiency while balancing the preservation or restoration of ecosystems with human's domestic, agricultural and industrial needs. Article 26 highlights the need to improve allocation efficiency of water within and across sectors for sustainable social and economic development. It also encourages the need to maximise the efficiency of human and financial resources in managing water resources.[94]

Improving water efficiency enables countries to increase resiliency to the impacts of water scarcity and maximise the benefits of existing water infrastructure.[95] Improved efficiency provides users with additional sources of water and reduces environmental degradation. In addition, the efficient use of water and related resources such as technological and financial resources maximises the economic and social welfare of water resources. Instead of increasing water supply in areas facing drought and scarcity, the first step should be investigating ways of improving water efficiency through reallocation or reducing wastage.

3.3.6   Concepts of water efficiency

Overall, there are four main interrelated concepts of water efficiency: technical efficiency, productive efficiency, product choice efficiency and allocative efficiency. Technical efficiency is referred to as water-use efficiency and requires measures such as recycling and reusing water, improving user practices and ensuring water infrastructure functions remain efficient. At the local level technical efficiency entails activities such as considering household products that are more water efficient. Productive efficiency is an economic concept that deals with the need to maximise the value of an output in relation to a specific level of inputs. The difference between productive efficiency and technical efficiency is that productive efficiency measures inputs and outputs in terms of their value. For example, an urban system may have a high leakage rate reducing the system's technical efficiency; however, from a productive efficiency framework, the costs of reducing the leakage may outweigh other benefits such as increased revenue, public health improvements, etc.[96] Product choice efficiency usually means goods and services provided reflect consumer's preferences and their ability or willingness to pay for them. In the water sector, however, it is usually water professionals that decide on the quality of service and the type of water infrastructure that is most appropriate.

This usually leads to distortions such as the provision of high-quality services to a small number of users. To avoid this, either the range of services and technology options available can be provided to a wider population or users participate in the decision-making process related to these options. Allocative efficiency is about allocating water resources for the production of water products and services and the allocation of available water among competing users in a way that maximises the benefits from their use. In allocative efficiency, costs and benefits need to include economic as well as social and environmental aspects.[97]

3.3.7 Management instruments in IWRM

IWRM is about modifying human systems to encourage people to use water resources efficiently and sustainably. In IWRM, demand management is based around using water more efficiently. It involves balancing supply and demand and focuses on the better use of existing supplies or reducing excessive consumption before new supplies are developed.[98] This is achieved using a variety of management instruments. Water resources assessments are required for informed decision-making and involve collecting hydrological, demographic and socioeconomic data and the setting up of routine data assembly and reporting. Water resources assessments are also important for mitigating floods and droughts. Assessments can be used for planning development options, resource use and human interactions. Communication and information instruments encourage a water-oriented society. Information is an important tool for changing behaviour through public awareness campaigns, school curricula, university water courses and professional training. Transparency of water resources data and product labelling of water-efficient appliances and practices are other key social change instruments. Having a conflict resolution mechanism in place is vital as conflict is endemic in the management of water resources in many places. Therefore, dispute resolution tools must be in place for users. Regulatory instruments are frequently used in the management of water and involve setting allocation and water-use limits. Regulation in this sense usually covers pollution control, service provision and land use. Regulatory instruments are frequently combined with economic, financial and technological instruments such as pricing, subsidies and other market tools to provide incentives for all water users to conserve water and use it efficiently and avoid pollution.[99] Technological instruments promote water efficiency and financing of IWRM projects. A summary of these management instrument is given in Table 3.4.

3.4 Framework for managing urban water sustainably: Integrated urban water management

IUWM has arisen from the many challenges to urban water resources including environmental degradation, rapidly growing urban populations and the impacts of climate change.[100] The key difference between IWRM and IUWM is the spatial

Table 3.4 Management instruments in IWRM

Instruments	Description
Water resources assessments	Data collection networks and assessment techniques
	Environmental impact assessments
	Risk management tools for flood and droughts
Communication and information	Raise awareness of the need for water conservation
	Informed stakeholder participation
Allocation and conflict resolution	Allocation of water resources through market instruments
	Allocation based on the valuation of costs and benefits
	Tools for conflict resolution: upstream versus downstream, sector versus sector, human versus nature
Regulatory	Direct controls – regulations, land use plans
	Economic – prices, tariffs, subsidies, fees, taxes
	Self-regulation – transparent benchmarking, product labelling
Technological	Research and development
	Efficiency guidelines
	Improving water supply infrastructure
Financing	Investment in IWRM by users, governments, private sector and donors

GWP. 2011. *IWRM – at a glance* [Online]. Available: http://www.gwp.org/Global/The%20Challenge/Resource%20material/IWRM%20at%20a%20glance.pdf (accessed 17 May 2016)

scale and the sector of application: IWRM is at the river basin level (which can include urban areas), while IUWM can be viewed as a subset of IWRM and is concerned with the management of water supply, wastewater and stormwater in urban areas.[101] In urban centres, IUWM advances both technological solutions for water management while simultaneously modifying the attitudes and behaviour of individuals and society towards scarce water resources.[102,103] IUWM can be applied by water managers in order to minimise the urban area's environmental impact on the surrounding environment.[104]

3.4.1   IUWM maximising pillars of sustainability

IUWM recognises actions that improve urban water systems extend beyond improving water quality and managing quantity. In particular, IUWM integrates the elements of the urban water cycle (water supply, sanitation, stormwater management and waste management) into both the city's urban development process and the management of the river basin the city is located in for the purpose of maximising water's many environmental, economic and social benefits equitably.[105,106,107,108,109,110] IUWM activities to maximise these benefits include improving water supply and consumption efficiency; ensuring adequate drinking water quality and wastewater treatment; improving economic efficiency of services to sustain operations and investments for water, wastewater and stormwater management; utilising alternative water sources; engaging communities in the decision-making process of water resources management; establishing and promoting water conservation programmes; and supporting capacity development of personnel and institutions that engage in IUWM.[111]

3.4.2   IUWM: Balancing demand for water with supply

In the context of IUWM, demand management comprises a set of policies that promote the better use of existing urban water supplies before plans are made to increase supply.[112,113] Specifically, demand management promotes water conservation, during times of both normal conditions and uncertainty, through changes in practices, cultures and attitudes of society towards water resources.[114,115] In addition to the environmental benefits of preserving ecosystems and their habitats, demand management is cost-effective compared to supply-side management as it allows the more efficient allocation of scarce financial resources (which would otherwise be required to build expensive dams and water transfer schemes from one river basin to another).[116] Finally, demand management ensures the equitable use of water by all users (domestic, industry, recreational, electricity, agriculture, etc.).

3.4.3   IUWM: Introducing demand management

To reduce risks to urban water security from non-climatic and climatic drivers, urban water managers can conduct numerous demand management activities to balance demand with supply to achieve urban water security. Water managers can pursue efficiency and conservation by optimising the efficiency of the existing system and investing in efficiency conservation programmes to influence customer behaviour. By reducing leakage and implementing cost-effective conservation programmes, water managers can help sustain water supply from existing sources, postpone or eliminate the need to invest in costly supply development projects and return water to ground and surface water supplies. Water managers can develop a diverse supply portfolio including rainwater harvesting, greywater reuse and recycling of water. Water managers can regularly update their long-term plans to ensure a consistent level of service, factoring in climate change extremes and impacts of urbanisation. Water managers can ensure their future investments in the urban water system are aimed at securing supply from local resources under their control.[117] Water managers should strive to balance water withdrawals with returns over time, ensuring a reliable supply of water for their service area as well as support regional water availability for others and for ecosystems.[118] Overall, IUWM activities advance technological solutions for water management while simultaneously modifying the attitudes and behaviour of individuals and society towards scarce water resources; this is important given humans are 'part and parcel of the environment rather than its masters'.[119,120,121,122] The benefits of doing so are reduction in the cost of building new infrastructure (new supplies) and meeting ever more stringent ecological requirements as water becomes more scarce.[123,124,125,126,127]

3.5   Other frameworks for managing urban water sustainably

Other frameworks for managing urban water sustainably that are common internationally include water sensitive urban design (WSUD), low impact development (LID) and low impact urban design and development (LIUDD).

3.5.1   Water sensitive urban design

WSUD is an approach to urban planning and design that integrates the management of the total water cycle into the urban development process. WSUD includes:

- The integrated management of groundwater, surface runoff, including stormwater, drinking water and wastewater to protect water-related environmental, recreational and cultural values
- Storage, treatment and beneficial use of runoff
- Treatment and reuse of wastewater
- Using vegetation for treatment purposes, water-efficient landscaping and enhancing biodiversity
- Utilising water-saving measures within and outside domestic, commercial and institutional premises to minimise requirements for drinking and non-drinking water supplies[128]

3.5.2   Low impact development

LID is a method of land development that seeks to maintain the natural hydrological character of a site or region. The natural hydrology of an area – the movement of water through a watershed – is shaped by local conditions to form a balanced and efficient system. However, urban development including roads, parking lots and rooftops alter the natural hydrology of an area by increasing runoff and reducing infiltration. LID designs aim to minimise these changes through source control – retaining more water on the site where it falls. This is achieved by using green roofs, natural landscapes and permeable pavement and other porous materials in developments or redevelopments.[129] The benefit of LID is that it can be used to increase local water supplies by collecting water on-site for non-potable uses, for example, landscape irrigation and flushing toilets or infiltrating water into soil to recharge groundwater supplies.[130]

3.5.3   Low impact urban design and development

LIUDD is an integrated design and development process that focuses on avoiding, at little or no extra cost, a wide range of adverse effects of urban development on aquatic and terrestrial ecological integrity while allowing urbanisation at all densities. In the context of sustainable urban water management, LIUDD aims to keep the unwanted effects of resource use to a minimum and ensure contaminants are treated at source through a variety of activities including increasing water supply efficiency through low-flow appliances and devices; recycling water; using appropriate garden design and plant species selection to minimise irrigation; collecting and treating rainwater for non-potable uses and replanting riparian areas and reducing impervious surfaces.[131]

Notes

1. ELZEN, B. & WIECZOREK, A. 2005. Transitions towards sustainability through system innovation. *Technological Forecasting and Social Change*, 72, 651–661.
2. WORLD COMMISSION ON ENVIRONMENT. 1992. *Our Common Future*. Geneva: Centre for Our Common Future.
3. VALLANCE, S., PERKINS, H. C. & DIXON, J. E. 2011. What is social sustainability? A clarification of concepts. *Geoforum*, 42, 342–348.
4. LIEBERHERR-GARDIOL, F. 2008. Urban sustainability and governance: issues for the twenty-first century. *International Social Science Journal*, 59, 331–342.
5. VALLANCE, S., PERKINS, H. C. & DIXON, J. E. 2011. What is social sustainability? A clarification of concepts. *Geoforum*, 42, 342–348.
6. CURWELL, S. & COOPER, I. 1998. The implications of urban sustainability. *Building Research & Information*, 26, 17–28.
7. DAVIDSON, M. 2009. Social sustainability: a potential for politics? *Local Environment*, 14, 607–619.
8. LOCAL GOVERNMENT ASSOCIATION OF NSW. 2012. *Barriers and drivers to sustainability*. Available: http://www.lgnsw.org.au/files/imce-uploads/35/barriers-and-drivers-to-sustainability.pdf (accessed 17 May 2016).
9. SMITH, M., DE GROOT, D. & BERGKAMP, G. 2006. *Pay: Establishing Payments for Watershed Services*. Gland, Switzerland: IUCN.
10. FINCO, A. & NIJKAMP, P. 2001. Pathways to urban sustainability. *Journal of Environmental Policy and Planning*, 3, 289–302.
11. LIEBERHERR-GARDIOL, F. 2008. Urban sustainability and governance: issues for the twenty-first century. *International Social Science Journal*, 59, 331–342.
12. UNEP. *Global initiative for resource-efficient cities*. Available: http://www.unep.org/pdf/GI-REC_4pager.pdf (accessed 11 May 2016).
13. LIEBERHERR-GARDIOL, F. 2008. Urban sustainability and governance: issues for the twenty-first century. *International Social Science Journal*, 59, 331–342.
14. UNITED NATIONS. 2012. *State of the World's Cities 2012–2013: Prosperity of Cities*. Nairobi, Kenya: United Nations Publications.
15. REES, W. & WACKERNAGEL, M. 2008. Urban ecological footprints: why cities cannot be sustainable – and why they are a key to sustainability. *In:* MARZLUFF, J. M., SHULENBERGER, E., ENDLICHER, W., ALBERTI, M., BRADLEY, G., RYAN, C., SIMON, U. & ZUMBRUNNEN, C. (eds) *Urban Ecology: An International Perspective on the Interaction Between Humans and Nature*. Boston, MA: Springer US.
16. BITHAS, K. P. & CHRISTOFAKIS, M. 2006. Environmentally sustainable cities. Critical review and operational conditions. *Sustainable Development*, 14, 177–189.
17. CURWELL, S. & COOPER, I. 1998. The implications of urban sustainability. *Building Research & Information*, 26, 17–28.
18. HAUGHTON, G. 1999. Environmental justice and the sustainable city. *Journal of Planning Education and Research*, 18, 233–243.

19. FINCO, A. & NIJKAMP, P. 2001. Pathways to urban sustainability. *Journal of Environmental Policy and Planning*, 3, 289–302.

20. REES, W. & WACKERNAGEL, M. 2012. Urban ecological footprints: why cities cannot be sustainable – and why they are a key to sustainability. *In*: MARZLUFF, J. M., SHULENBERGER, E., ENDLICHER, W., ALBERTI, M., BRADLEY, G., RYAN, C., SIMON, U. & ZUMBRUNNEN, C. (eds) *Urban Ecology: An International Perspective on the Interaction Between Humans and Nature*. Boston, MA: Springer US.

21. HAUGHTON, G. 1999. Environmental justice and the sustainable city. *Journal of Planning Education and Research*, 18, 233–243.

22. CURWELL, S. & COOPER, I. 1998. The implications of urban sustainability. *Building Research & Information*, 26, 17–28.

23. JACKSON, T. 2005. *Motivating sustainable consumption: a review of evidence on consumer behaviour and behavioural change: a report to the Sustainable Development Research Network*. London: Centre for Environmental Strategy, University of Surrey.

24. SATTERTHWAITE, D. 2007. *The Transition to a Predominantly Urban World and Its Underpinnings*. London: Human Settlements Programme, IIED.

25. NEUMAYER, E. 2012. Human development and sustainability. *Journal of Human Development and Capabilities*, 13, 561–579.

26. MILBRATH, L. W. 1995. Psychological, cultural, and informational barriers to sustainability. *Journal of Social Issues*, 51, 101–120.

27. BARBIER, E. 2011. The policy challenges for green economy and sustainable economic development. *Natural Resources Forum*, 35, 233–245.

28. NEUMAYER, E. 2012. Human development and sustainability. *Journal of Human Development and Capabilities*, 13, 561–579.

29. HAUGHTON, G. 1999. Environmental justice and the sustainable city. *Journal of Planning Education and Research*, 18, 233–243.

30. SPENCE, A. & PIDGEON, N. 2009. Psychology, climate change and sustainable behaviour. *Environment: Science and Policy for Sustainable Development*, 51, 8–18.

31. PIKE, C., DOPPELT, B. & HERR, M. 2010. Climate communications and behavior change: a guide for practitioners. *The Climate Leadership Initiative* [Online]. Available: https://www.seek.state.mn.us/resource/climate-communications-and-behavior-change-guide-practitioners (accessed 2 June 2016).

32. BARBIER, E. 2011. The policy challenges for green economy and sustainable economic development. *Natural Resources Forum*, 35, 233–245.

33. DARNAULT, C. J. G. 2008. Sustainable development and integrated management of water resources. *Overexploitation and Contamination of Shared Groundwater Resources*. Dordrecht: Springer.

34. BARBIER, E. 2011. The policy challenges for green economy and sustainable economic development. *Natural Resources Forum*, 35, 233–245.

35. DASGUPTA, P. 2008. Nature in economics. *Environmental and Resource Economics*, 39, 1–7.

36. BARBIER, E. 2011. The policy challenges for green economy and sustainable economic development. *Natural Resources Forum*, 35, 233–245.

37. NEUMAYER, E. 2012. Human development and sustainability. *Journal of Human Development and Capabilities*, 13, 561–579.

38. GOODLAND, R. 1995. The concept of environmental sustainability. *Annual Review of Ecology and Systematics*, 26, 1–24.

39. CURWELL, S. & COOPER, I. 1998. The implications of urban sustainability. *Building Research & Information*, 26, 17–28.

40. SALLES, J.-M. 2011. Valuing biodiversity and ecosystem services: why put economic values on Nature? *Comptes Rendus Biologies*, 334, 469–482.

41. GOLDMAN, R. L. 2010. Ecosystem services: how people benefit from nature. *Environment*, 52, 15–23.

42. Ibid.

43. SALLES, J.-M. 2011. Valuing biodiversity and ecosystem services: why put economic values on Nature? *Comptes Rendus Biologies*, 334, 469–482.

44. BOYER, T. & POLASKY, S. 2004. Valuing urban wetlands: a review of non-market valuation studies. *Wetlands*, 24, 744–755.

45. GOLDMAN, R. L. 2010. Ecosystem services: how people benefit from nature. *Environment*, 52, 15–23.

46. EDWARDS, P. J. & ABIVARDI, C. 1998. The value of biodiversity: where ecology and economy blend. *Biological Conservation*, 83, 239–246.

47. BOYER, T. & POLASKY, S. 2004. Valuing urban wetlands: a review of non-market valuation studies. *Wetlands*, 24, 744–755.

48. Ibid.

49. Ibid.

50. PIKE, C., DOPPELT, B. & HERR, M. 2010. Climate communications and behavior change: a guide for practitioners. *The Climate Leadership Initiative* [Online]. Available: https://www.seek.state.mn.us/resource/climate-communications-and-behavior-change-guide-practitioners (accessed 2 June 2016).

51. FINCO, A. & NIJKAMP, P. 2001. Pathways to urban sustainability. *Journal of Environmental Policy and Planning*, 3, 289–302.

52. HAUGHTON, G. 1999. Environmental justice and the sustainable city. *Journal of Planning Education and Research*, 18, 233–243.

53. VALLANCE, S., PERKINS, H. C. & DIXON, J. E. 2011. What is social sustainability? A clarification of concepts. *Geoforum*, 42, 342–348.

54. HAUGHTON, G. 1999. Environmental justice and the sustainable city. *Journal of Planning Education and Research*, 18, 233–243.

55. CURWELL, S. & COOPER, I. 1998. The implications of urban sustainability. *Building Research & Information*, 26, 17–28.

56. LIEBERHERR-GARDIOL, F. 2008. Urban sustainability and governance: issues for the twenty-first century. *International Social Science Journal*, 59, 331–342.

57. CUTHILL, M. 2010. Strengthening the 'social' in sustainable development: Developing a conceptual framework for social sustainability in a rapid urban growth region in Australia. *Sustainable Development*, 18, 362–373.

58. LIEBERHERR-GARDIOL, F. 2008. Urban sustainability and governance: issues for the twenty-first century. *International Social Science Journal*, 59, 331–342.

59. BUTLER, D., FARMANI, R., FU, G., WARD, S., DIAO, K. & ASTARAIE-IMANI, M. 2014. A new approach to urban water management: safe and sure. *Procedia Engineering*, 89, 347–354.

60. ADB. 2014. *Urban climate change resilience: a synopsis*. Available: http://www.adb.org/publications/urban-climate-change-resilience-synopsis (accessed 17 May 2016).

61. WORLD BANK. 2012. Integrated urban water management: a summary note. *Blue Water Green Cities* [Online]. Available: http://siteresources.worldbank.org/INTLAC/Resources/257803-1351801841279/1PrincipalIntegratedUrbanWaterManagementENG.pdf (accessed 17 May 2016).

62. ENGEL, K. 2011. *Big cities. Big water. Big challenges: water in an urbanizing world*. Available: http://www.wwf.se/source.php/1390895/Big%20Cities_Big%20Water_Big%20Challenges_2011.pdf (accessed 17 May 2016).

63. WORLD BANK. 2012. Integrated urban water management: a summary note. *Blue Water Green Cities* [Online]. Available: http://siteresources.worldbank.org/INTLAC/Resources/257803-1351801841279/1PrincipalIntegratedUrbanWaterManagementENG.pdf (accessed 17 May 2016).

64. Ibid.

65. JOWSEY, E. 2012. The changing status of water as a natural resource. *International Journal of Sustainable Development & World Ecology*, 19, 433–441.

66. JØNCH-CLAUSEN, T. & FUGL, J. 2001. Firming up the conceptual basis of Integrated Water Resources Management. *International Journal of Water Resources Development*, 17, 501–510.

67. UN-WATER. 2013. *Water security and the global water agenda.* Available: http://www. unwater.org/downloads/watersecurity_analyticalbrief.pdf (accessed 17 May 2016).

68. LOUCKS, D. P. 2000. Sustainable water resources management. *Water International*, 25, 3–10.

69. OFFERMANS, A., HAASNOOT, M. & VALKERING, P. 2011. A method to explore social response for sustainable water management strategies under changing conditions. *Sustainable Development*, 19, 312–324.

70. JOWSEY, E. 2012. The changing status of water as a natural resource. *International Journal of Sustainable Development & World Ecology*, 19, 433–441.

71. UN-WATER. 2013. *Water security and the global water agenda.* Available: http://www. unwater.org/downloads/watersecurity_analyticalbrief.pdf (accessed 17 May 2016).

72. SMITH, M., DE GROOT, D. & BERGKAMP, G. 2006. *Pay: Establishing Payments for Watershed Services.* Gland: IUCN.

73. Ibid.

74. OECD. 2010. *Pricing Water Resources and Water and Sanitation Services*, Paris: OECD Publishing.

75. JØNCH-CLAUSEN, T. & FUGL, J. 2001. Firming up the conceptual basis of Integrated Water Resources Management. *International Journal of Water Resources Development*, 17, 501–510.

76. VAN DER ZAAG, P. & SAVENIJE, H. 2006. *Water as an Economic Good: The Value of Pricing and the Failure of Markets.* Delft: UNESCO-IHE.

77. CORFEE-MORLOT, J., KAMAL-CHAOUI, L., DONOVAN, M., COCHRAN, I., ROBERT, A. & TEASDALE, P.-J. 2009. *Cities, Climate Change and Multilevel Governance.* Paris: OECD Publishing.

78. BITHAS, K. 2008. The sustainable residential water use: sustainability, efficiency and social equity. The European experience. *Ecological Economics*, 68, 221–229.

79. JØNCH-CLAUSEN, T. & FUGL, J. 2001. Firming up the conceptual basis of Integrated Water Resources Management. *International Journal of Water Resources Development*, 17, 501–510.

80. MEDEMA, W., MCINTOSH, B. S. & JEFFREY, P. J. 2008. From premise to practice: a critical assessment of integrated water resources management and adaptive management approaches in the water sector. *Ecology and Society*, 13, 1–18.

81. Ibid.

82. GLOBAL WATER PARTNERSHIP. 2011. *What is IWRM?* [Online]. Available: http:// www.gwp.org/en/The-Challenge/What-is-IWRM/ (accessed 17 May 2016).

83. Ibid.

84. Ibid.

85. DAVIE, T. 2008. *Fundamentals of Hydrology.* London: Taylor & Francis.

86. GLOBAL WATER PARTNERSHIP. 2011. *What is IWRM?* [Online]. Available: http:// www.gwp.org/en/The-Challenge/What-is-IWRM/ (accessed 17 May 2016).

87. UN DOCUMENTS. 2011. *The Dublin Statement on Water and Sustainable Development* [Online]. Available: http://www.un-documents.net/h2o-dub.htm (accessed 17 May 2016).

88. Ibid.

89. Ibid.

90. Ibid.

91. Ibid.

92. UN. 1992. *Agenda 21* [Online]. Available: http://www.un.org/esa/dsd/agenda21/ (accessed 17 May 2016).

93. Ibid.

94. UN DEPARTMENT OF ECONOMIC AND SOCIAL AFFAIRS: DIVISION FOR SUSTAINABLE DEVELOPMENT. 2011. *World Summit on Sustainable Development Johannesburg Plan of Implementation* [Online]. Available: http://www.un.org/esa/ sustdev/documents/WSSD_POI_PD/English/POIChapter4.htm (accessed 17 May 2016).

95. WHO. 2009. *Vision 2030: the resilience of water supply and sanitation in the face of climate change.* Available: http://www.who.int/water_sanitation_health/publications/ 9789241598422/en/ (accessed 17 May 2016).

96. GWP. 2005. *Taking an integrated approach to improving water efficiency.* Available: http://www.gwp.org/en/ToolBox/PUBLICATIONS/Technical-Briefs/ (accessed 2 June 2016).

97. Ibid.

98. GWP (ed.) 2004. *Catalyzing Change: A Handbook for Developing Integrated Water Resource Management Strategies (IWRM) and Water Efficiency Strategies.* Stockholm: Global Water Partnership Technical Committee with support from Norway's Ministry of Foreign Affairs.

99. Ibid.

100. BROWN, R., KEATH, N. & WONG, T. 2009. Urban water management in cities: historical, current and future regimes. *Water Science & Technology,* 59(5), 847–855.

101. MAHEEPALA, S., BLACKMORE, J., DIAPER, C., MOGLIA, M., SHARMA, A. & KENWAY, S. 2010. *Towards the adoption of integrated urban water management approach for planning.* Conference Proceedings of the Water Environment Federation, WEFTEC 2010: Session 91 through to 100, pp. 6734–6753.

102. BAHRI, A. 2012. *Integrated urban water management.* Available: http://www.gwp.org/ Global/The%20Challenge/Resource%20material/GWP_TEC16.pdf (accessed 17 May 2016).

103. LOUCKS, D. P. 2000. Sustainable water resources management. *Water International,* 25, 3–10.

104. MAHEEPALA, S., BLACKMORE, J., DIAPER, C., MOGLIA, M., SHARMA, A. & KENWAY, S. 2010. *Towards the adoption of integrated urban water management approach for planning.* Conference Proceedings of the Water Environment Federation, WEFTEC 2010: Session 91 through to 100, pp. 6734–6753.

105. GABE, J., TROWSDALE, S. & VALE, R. 2009. Achieving integrated urban water management: planning top-down or bottom-up? *Water Science and Technology,* 59, 1999–2008.

106. WORLD BANK. 2012. Integrated urban water management: a summary note. *Blue Water Green Cities* [Online]. Available: http://siteresources.worldbank.org/INTLAC/ Resources/257803-1351801841279/1PrincipalIntegratedUrbanWaterManagementENG.pdf (accessed 17 May 2016).

107. MAHEEPALA, S. & BLACKMORE, J. 2008. Integrated urban water management. In: *Transitions: Pathways Towards Sustainable Urban Development in Australia.* Chapter 30, Collingwood, Victoria: CSIRO Publishing, pp. 568–588.

108. GABE, J., TROWSDALE, S. & VALE, R. 2009. Achieving integrated urban water management: planning top-down or bottom-up? *Water Science and Technology,* 59, 1999–2008.

109. VAN DE MEENE, S., BROWN, R. & FARRELLY, M. 2011. Towards understanding governance for sustainable urban water management. *Global Environmental Change,* 21, 1117–1127.

110. BAHRI, A. 2012. *Integrated urban water management.* Available: http://www.gwp. org/Global/The%20Challenge/Resource%20material/GWP_TEC16.pdf (accessed 17 May 2016).

111. UNEP. 2011. *Integrated urban water management* [Online]. Available: http://www.unep.or.jp/ietc/brochures/iuwm.pdf (accessed 7 June 2016).

112. GLOBAL WATER PARTNERSHIP. 2012. *Water Demand Management (WDM) – the Mediterranean experience*. Technical focus paper [Online]. Available: http://www.gwp.org/en/gwp-in-action/News-and-Activities/Global-Water-Partnership-launches-new-publications-at-World-Water-Week-2012/ (accessed 2 June 2016).

113. SAVENIJE, H. & VAN DER ZAAG, P. 2002. Water as an economic good and demand management: paradigms with pitfalls. *Water International*, 27, 98–104.

114. Ibid.

115. GLOBAL WATER PARTNERSHIP. 2012. *Water Demand Management (WDM) – the Mediterranean experience*. Technical focus paper [Online]. Available: http://www.gwp.org/en/gwp-in-action/News-and-Activities/Global-Water-Partnership-launches-new-publications-at-World-Water-Week-2012/ (accessed 2 June 2016).

116. Ibid.

117. THE JOHNSON FOUNDATION. 2014. *Charting new waters: ensuring urban water security in water-scarce regions of the United States.* Available: http://www.johnsonfdn.org/sites/default/files/conferences/whitepapers/14/05/19/cnw_urbanwatersecuritymay2014.pdf (accessed 17 May 2016).

118. UN-WATER. 2014. *The World Water Development Report 2014: water and energy.* Available: http://www.unwater.org/publications/publications-detail/en/c/218614/ (accessed 2 June 2016).

119. MOLLE, F. 2009. Water and society: new problems faced, new skills needed. *Irrigation and Drainage*, 58, S205-S211.

120. VAN DER BRUGGE, R., ROTMANS, J. & LOORBACH, D. 2005. The transition in Dutch water management. *Regional Environmental Change*, 5, 164–176.

121. BAHRI, A. 2012. *Integrated urban water management.* Available: http://www.gwp.org/Global/The%20Challenge/Resource%20material/GWP_TEC16.pdf (accessed 17 May 2016).

122. LOUCKS, D. P. 2000. Sustainable water resources management. *Water International*, 25, 3–10.

123. Ibid.

124. VAN DE MEENE, S., BROWN, R. & FARRELLY, M. 2011. Towards understanding governance for sustainable urban water management. *Global Environmental Change*, 21, 1117–1127.

125. WORLD BANK. 2012. Integrated urban water management: a summary note. *Blue Water Green Cities* [Online]. Available: http://siteresources.worldbank.org/INTLAC/Resources/257803-1351801841279/1PrincipalIntegratedUrbanWaterManagementENG.pdf (accessed 17 May 2016).

126. MAHEEPALA, S., BLACKMORE, J., DIAPER, C., MOGLIA, M., SHARMA, A. & KENWAY, S. 2010. Towards the adoption of integrated urban water management approach for planning. *Proceedings of the Water Environment Federation*, 2010, 6734–6753.

127. UNEP. 2011. *Integrated urban water management* [Online]. Available: http://www.unep.or.jp/ietc/brochures/iuwm.pdf (accessed 17 May 2016).

128. GOVERNMENT OF SOUTH AUSTRALIA. 2010. *Water sensitive urban design – Greater Adelaide region technical manual.* Available: https://www.sa.gov.au/topics/housing-property-and-land/building-and-development/land-supply-and-planning-system/water-sensitive-urban-design (accessed 17 May 2016).

129. CALIFORNIA WATER AND LAND USE PARTNERSHIP. 2006. *Diagram adapted from Prince George's County Maryland Low-Impact Development Design Strategies Low*

Impact Development (LID) A Sensible Approach to Land Development and Stormwater Management. Available: http://www.coastal.ca.gov/nps/lid-factsheet.pdf (accessed 17 May 2016).

130. NRDC. 2009. *Water saving solutions: stopping pollution at its source with low impact development.* Available: https://www.nrdc.org/water/lid/files/flid.pdf (accessed 17 May 2016).

131. ROON, M. V. R. A. H. V. 2010. *Low impact urban design and development: the big picture.* Available: http://www.mwpress.co.nz/__data/assets/pdf_file/0006/70494/LRSS_37_LIUDD_big_picture-.pdf (accessed 17 May 2016).

4 Demand management to achieve urban water security

Introduction

In IUWM, demand management involves the better use of existing water supplies before plans are made to further increase supply. In particular, demand management promotes water conservation during times of both normal and atypical conditions, through changes in practices, culture and people's attitudes towards water resources. Demand management involves communicating ideas, norms and innovating methods for water conservation across individuals and society; the purpose of demand management is to positively adapt society to reduce water consumption patterns and achieve water security.

This chapter first discusses the purpose of demand management strategies, before discussing two types of demand management strategies and instruments available to urban water managers. The chapter then discusses the various demand management tools available to water managers. Finally, the chapter discusses how urban water managers need to develop portfolios of demand management tools to achieve urban water security.

4.1   Purpose of demand management

Demand management comprises a set of policies that promote the better use of existing urban water supplies before plans are made to increase supply.[1,2] Specifically, demand management promotes water conservation, during times of both normal conditions and uncertainty, through changes in practices, cultures

Urban Water Security, First Edition. Robert C. Brears.
© 2017 John Wiley & Sons, Ltd. Published 2017 by John Wiley & Sons, Ltd.

and people's attitudes towards water resources.[3,4] In addition to the environmental benefits of preserving ecosystems and their habitats, demand management is cost-effective compared to supply-side management as it allows the more efficient allocation of scarce financial resources (which would otherwise be required to build expensive dams and water transfer schemes from one river basin to another).[5] Finally, demand management ensures the equitable use of water by all users (domestic, industry, recreational, electricity, agriculture, nature, etc.).

With regard to actual water resources, demand management seeks to reduce the loss and misuse in various water sectors (intra-sector efficiency); optimise water use by assuring a reasonable allocation between various users (cross-sectoral efficiency) while taking into account the supply needs of downstream ecosystems and *in situ* uses of water such as recreational, fisheries, agricultural and energy production; facilitate major financial and infrastructural savings for countries, cities, companies and users by minimising the need to meet increasing demand with new water supplies; and reduce the stress on water resources by reducing or halting unsustainable exploitation of water resources.[6] The benefits of which are listed in Figure 4.1.

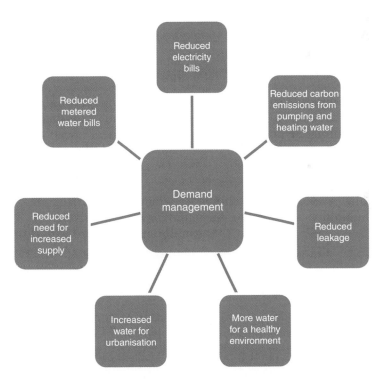

Figure 4.1 Benefits of demand management (ECONOMIC AND SOCIAL RESEARCH COUNCIL. 2008. *Behavioural change and water efficiency*. Available: http://webarchive. nationalarchives.gov.uk/20080821115857/http://esrc.ac.uk/ESRCInfoCentre/about/CI/ events/esrcseminar/BehaviouralChangeandWaterEfficiency.aspx?ComponentId=25751& SourcePageId=6066 (accessed 9 May 2016)).

4.1.1 Types of demand management strategies and instruments

Water managers can use two types of demand management strategies to influence the norms and values of society towards water resources: antecedent and consequential strategies. Antecedent strategies attempt to influence the determinants of target behaviour through activities such as increasing individual's knowledge or awareness of problems through information campaigns, behavioural commitments and prompting, the assumption being these strategies can influence the determinants of behaviour before its performance. Consequential strategies (feedback, rewards and punishments) are all assumed to influence the determinants of target behaviour *after* the performance of the behaviour. The latter strategy assumes that feedback, both positive and negative, of the consequences of that behaviour influences the likelihood of that behaviour being performed in the future.[7] To implement these strategies, water managers can use two types of demand management instruments to influence the norms and values of society towards water resources: regulatory and technological instruments and communication and information instruments.

4.2 Regulatory and technological demand management instruments

Regulatory and technological instruments are frequently used in the management of water and involve setting allocation and water-use limits. In addition, regulatory and technological instruments are used to provide incentives for all water users to conserve water and use it efficiently.

4.2.1 Pricing of water

In water resource management, economic theory suggests that demand for water should behave like any other goods – as price increases, water use decreases. As such, pricing of water is commonly used as an incentive to promote water conservation.[8] In water conservation, the pricing of water is used to internalise the environmental costs of consumption decisions; serve as an allocation mechanism directing water where it is most valued, protecting water resources from wasteful use; and promote water efficiency (doing more with the same amount of water), for example, pricing of water can incentivise the development or adoption of water-saving technologies.[9,10,11,12]

In using price as a mechanism to promote water conservation, water managers can use a variety of price structures, all of which send to individuals and communities different conservation signals. Specifically, there are three dimensions to pricing of water available to water managers: first, whether the tariff is directly linked to water consumption; second, whether it covers fixed and variable costs; and, third, whether the tariff level changes as more water is consumed, where change in price occurs in differing blocks of water consumed.[13]

Flat rate

A flat rate is essentially a fixed charge for water usage regardless of the volume used, where typically the size of the fixed charge is related to the customer's property value.[14,15] While fixed prices enable water utilities to raise sufficient revenue for the operation and maintenance of the water supply network, it does not provide any incentive for individuals and communities to conserve water.[16,17]

Volumetric rate

A volumetric rate is a charge based on the volume of water used at a constant rate, for example, $1 per cubic metre of water used. Therefore, the amount users pay for water is strictly based on the amount of water consumed.[18]

Increasing block tariff

An increasing block tariff contains different prices for two or more prespecified blocks of water, with price increasing with each successive block. The water supplier must therefore decide on: first, the number of blocks; second, the volume of water use associated with each block; and, third, the price charged for each block with decision one being a management decision, while decisions two and three are political and social decisions.[19] With an increasing block tariff, water suppliers can provide a social net for low-income households by providing a low price for a specified amount of water and a higher price for amounts above this minimum volume.[20] An increasing block tariff system is frequently viewed as the best compromise between economic efficiency and social equity for domestic water supply, where the most essential uses of water for drinking, cooking and sanitation are priced lower than the lowest value uses of water such as watering lawns and washing cars.[21]

Fixed and variable pricing

The pricing of water can include a two-part tariff system: a fixed and variable component. In the fixed component, water users pay one amount independently of consumption to cover administrative and infrastructural costs of supplying water. Meanwhile, the variable amount is based on the amount of water consumed and covers the costs of pumping, treating and distributing water.[22] It is also possible for the variable part to be designed with increasing block rates.[23] The advantage of a two-part tariff system is that it stabilises the revenue base of the supplier; specifically, the fixed component protects the supplier from demand fluctuations, reducing financial risk, while the variable component charges consumers according to their consumption levels, therefore encouraging conservation.[24]

Water Corporation of Western Australia's block pricing structure

CASE 4.1

The Water Corporation of WA has a pricing structure where the price of water per kilolitre increases as customers use more water across the billing year (Table 4.1). There are six billing periods with water use accruing across the year resulting in water becoming more expensive towards the end of the billing year as customers move into the higher water-use bracket. The purpose of this is to encourage the careful use of water.

Table 4.1 Water Corporation's pricing structure of water

Price	Total water usage (kilolitre)
$1.518	0–150
$2.023	151–500
$2.864	500+

WATER CORPORATION. 2015b. *Your bill and charges* [Online]. Available: http://www.watercorporation.com.au/my-account/your-bill-and-charges (accessed 17 May 2016)

Sydney Water's fixed and variable water pricing structure

CASE 4.2

Sydney Water charges all its domestic customers a fixed water service charge for connecting to the water supply, a fixed wastewater (sewage) service charge, a fixed storm water service charge and a volumetric water usage charge for the amount of water used (Table 4.2).

Table 4.2 Sydney Water's pricing structure

Water price	Description	2015–2016 charge
Water service charge	If you have your own meter	$25.64 a quarter
	If you share a meter	$25.64 a quarter
	If you don't have a meter	$128.11 a quarter
Wastewater service charge	Your wastewater service charge	$152.29 a quarter
Storm water service charge	If you live in a house	$21.52 a quarter
	If you live in a unit	$7.90 a quarter
Water usage charge	Your water usage	$2.276 a kilolitre

SYDNEY WATER. 2015b. *Our prices* [Online]. Available: https://www.sydneywater.com.au/SW/accounts-billing/understanding-your-bill/our-prices/index.htm (accessed 17 May 2016)

Seattle's seasonal drinking water rates

Seattle charges its residential customers a peak residential water rate from May 16 to September 15 each year with peak rates incorporating a three-tiered rate structure with progressively higher rates as water consumption increases (Table 4.3).

Table 4.3 Seattle's seasonal pricing structure

Water usage	Consumption volume	Tariff (per ccf) (one ccf = 100 cubic feet or 748 gallons of water)
First tier	Up to 10 ccf in 60 days	$5.13
Second tier	Next 26 ccf in 60 days	$6.34
Third tier	Over 36 ccf in 60 days	$11.80

SEATTLE PUBLIC UTILITIES. 2015. *Third-tier water rates* [Online]. Available: http://www.seattle.gov/util/MyServices/Rates/WaterRates/ThirdTierWaterRates/index.htm (accessed 17 May 2016)

Seasonal rates

Changes in water-use patterns from season to season due to changes in weather occur in many systems. If these fluctuations are extreme, water utilities often implement a seasonal rate structure. Under this rate structure, customers are charged a higher rate during peak season. Utilities can apply one of two forms of seasonal rate structure. The first option is to set one rate for the off-peak season and one for the peak season (e.g. these rates can be volumetric or with increasing block tariffs). The second option is to charge one rate (volumetric or increasing block rate) and apply excess usage charges, that is, charge for water use in excess of that used on average during off-peak times. Seasonal rates can encourage conservation, reduce peak use and therefore limit the need to expand system capacity.[25]

4.2.2 What is the right price?

In traditional water resource management, water was viewed as cheap and plentiful. As such, water prices were set to achieve only partial cost recovery, with a rate of return on water assets frequently below commercial levels.[26] This 'underpricing' of water is frequently justified on the basis of egalitarianism, where water is a natural asset indispensable for human survival, therefore should be priced low for low-income consumers to ensure social equity.[27] However, water is no longer a natural asset. Instead, its current use in urban areas requires transporting, treating and distributing, resulting in substantial (environmental and economic) costs of providing that water.[28] Nevertheless, these costs are seldom fully factored into the price of water, leading to the frequent underpricing of water in cities, which promotes overconsumption, further intensifying water scarcity, which in turn further raises the costs of using water resources.[29]

This chain of events adversely impacts the sustainable use of water in two ways: first, the actual total amount of water used exceeds sustainable levels, creating welfare losses for current and future generations; and, second, the increasing costs induced by an increase in scarcity leads to further welfare losses as water used for numerous economic and social activities needs to be restricted.[30] As a result of egalitarian underpricing of water, low-income households will be more intensely affected in the future by rising water prices due to increased scarcity; because any price increase will be greater than today's full economic cost of providing water, therefore, future prices will take a greater percentage of household income compared with more wealthier households. As such, water prices need to reflect the full costs of providing water including environmental, social and supply costs to ensure efficient and sustainable use prevails in order to avoid future price increases.[31] Nonetheless, the price of water should be based on a 'reasonable price structure' that not only aims at cost recovery but also takes into account access to safe water for the poor and the state of the natural environment from which water is drawn from. Therefore, giving a reasonable price sends a clear signal to water users that water should be used wisely.[32,33]

Price elasticity

With ordinary economic goods there is a relation between price and quantity demanded (demand curve). The slope of the demand curve is called the price elasticity of demand (E), defined as the percentage decrease in demand resulting from an increase in price. The elasticity is a negative number since demand is expected to decrease as price increases. For water, E usually ranges from −1 to 0. However, elasticity is not a constant value as it depends on price and type of water use.[34] Water has a 'special characteristic' in that as water use becomes more essential for health and sanitation, the more elasticity becomes rigid (E is close to 0). For instance, people need drinking water to survive and will pay any price for it. The less essential the water use is for survival (e.g. watering lawns and washing cars), the more elastic the demand becomes (E becomes closer to −1). This is because higher efficiencies can be achieved through water conservation and water-saving technologies.[35]

Objectives of pricing

The most common objective of pricing is to raise revenue in a politically acceptable way rather than modify behaviour.[36] Nonetheless, there are numerous additional objectives that must be considered when setting the price for scarce water resources, which are listed in Table 4.4.

4.2.3 Water meters

Before water users can be charged for the amount of water consumed, the dwelling or building must have water meters to measure the volume of water consumed. In addition to being able to charge for volume consumed, water meters provide

Table 4.4 Objectives of pricing

Objectives of tariffs	Description
Economic efficiency	Prices should cover the full costs of providing water (environmental, social and supply costs)
Environmental sustainability	Economic instruments should be concerned about attaining/ensuring the sustainable use of water resources and the health of the environment for present and future generations
Fairness	Frequently it is argued that rising block prices for water is the fairest way of pricing water as it enables households to receive a certain minimum amount of water at a low price; however, it raises equity questions on what is the minimum amount people should receive before they pay a higher tariff. In addition, it promotes a loss of sense in the value of water. In particular, if the block is too large, then average consumers will fit nearly all their consumption within the first allowance, and if it is too small, then it will impact low-income households unfairly
Equity	Prices should treat water users equally – all who purchase water at the same costs should pay the same price
Revenue sufficiency	The total revenue collected from pricing water meets all the costs of providing the water, and it enables providers to invest in infrastructure as well as water conservation programmes
Net income stability	Prices should be designed so it minimises changes in net revenue due to unexpected fluctuations in demand caused by economic or weather conditions
Simplicity and understanding	Prices should be readily understandable to water users and others who are expected to make decisions based on water prices
Resource conservation	Prices should promote the conservation of scarce water resources
Enforceable	Prices should be simple to enforce
Simple	Prices should be simple to administer and simple to understand

BOLAND, J. J. 1993. Pricing urban water: principles and compromises. *Water Resources Update*, 92, 7–10

ECONOMIC AND SOCIAL RESEARCH COUNCIL. 2008. *Behavioural change and water efficiency*. Available: http://webarchive. nationalarchives.gov.uk/20080821115857/http://esrc.ac.uk/ESRCInfoCentre/about/CI/events/esrcseminar/ BehaviouralChangeandWaterEfficiency.aspx?ComponentId=25751&SourcePageId=6066 (accessed 9 May 2106)

ROGERS, P., DE SILVA, R. & BHATIA, R. 2002. Water is an economic good: how to use prices to promote equity, efficiency, and sustainability. *Water Policy*, 4, 1–17

OECD. 2012b. *Environmental Outlook to 2050: The Consequences of Inaction*. Paris: OECD

CAP-NET. 2008. *Economics in sustainable water management: training manual and facilitators' guide*. Available: http://www.euwi. net/files/Cap_net_EUWI_FWG_GWP_Manual_Economics_of_water_FINAL.pdf (accessed 9 May 2016)

numerous benefits including decreases in associated costs of heating water for use in showers, washing machines and so on; decreases in supply-side costs of infrastructure piping, storage, distribution and treatment; and decreases in environmental costs associated with water services and use including energy, water and carbon emissions.[37]

A study in the United Kingdom found water meters reduced household water consumption by 10 percent during normal climatic conditions and by 20–50 percent during hot/dry periods.[38] Nonetheless, there are numerous costs involved with the use of meters including installation costs; capital/operating costs of metering

Water metering in Calgary

In 2010, Calgary's city council mandated the metering of all flat-rate customers. Between 2010 and 2014, approximately 10 000 metres were installed on a community-by-community basis with the goal of achieving universal metering by 2014. Water meters will also be installed in all new homes built, when new accounts are created (a home is sold) and when customers voluntarily request a meter. This is part of the city's water efficiency goal of reducing water consumption by 30 percent over 30 years, ensuring Calgary can accommodate its population growth with the same amount of water from the river used in 2003.[42]

regimes including campaigning, administration, information, monitoring and maintenance costs; social costs associated with vulnerability of low-income households; and loss of benefits due to decreased water consumption by users.[39,40] For example, the same UK study found that on average it costs £200 per household to have a meter installed. In addition, the extra costs of providing metered services are on average £52 per annum, a third of the average water bill. Meanwhile, a 10 percent reduction in water consumption equates to a saving of only £15 per household per annum. As such, the case for metering depends on how much water can be conserved relative to the cost of metering.[41]

Smart meters

Automatic meter readers (AMRs) are 'one-way' automated meter readers that send usage data back to the utility. In comparison, advanced metering infrastructure (AMI) is a 'two-way' solution that creates a network between the meters and the utility's information system. Data flows both ways facilitating not only remote meter reading but also the ability to remotely activate meters and the use of variable pricing.[43]

Specifically, AMI/smart meters target the management and extraction of useful information from large amounts of high-resolution consumption data. The data can be used to develop customised awareness programmes to influence behaviour change. From the utility side smart meters provide many benefits including leakage detection/energy reduction, demand forecasting, awareness campaigns, promotion of efficiency appliances, water pricing/tariffs, case for investment and performance indicators. On the customer side smart meters can provide information on aspects including where/when is water used, comparison of own water use and against other customers, leakage and associated energy consumption. Overall, the specific applications of smart meters include comparing water consumption with other consumers (e.g. neighbour in the same building or street), comparing water consumption with standard profiles (consumers with the same socio-demographic factors), comparing household water consumption with most efficient users, comparing energy patterns associated with water use in the same household receiving information on specific and alternative pricing schemes, forecasting the next water bill and forecasting the component of the next energy bill associated with water consumption.[44]

Smart meter trial in Long Beach, California

The Long Beach Board of Water Commissioners approved a new programme that offered the first one hundred Long Beach residents to sign up to a free smart meter. The smart meters will help customers track their daily water consumption through a secure website enabling them to find new ways of conserving water. The smart meter is installed over a residence's existing water meter and collects water consumption data in five-minute increments. The data is downloaded to a website that each resident will be given private access to, allowing them to view their daily water use. The smart meters will also be offered to existing metered households and businesses that have received multiple water violation letters from the Long Beach Water Department.[45]

CASE 4.5

4.2.4   Reducing unaccounted-for water

A major issue for water utilities is addressing the considerable difference between the volume of water they treat and distribute and the volume of water invoiced to the customer. The gap is unaccounted-for water (UFW) and often amounts to between 25 and 50 percent of the total amount of water collected, treated and distributed. Three main categories of UFW are apparent losses, or commercial losses, caused by inaccurate customer metering, data-logging errors and illegal connections to the network; real losses, or physical losses, comprising leakage from all parts of the system; and overflow at the utility's storage tanks. Real losses are caused by poor operations and maintenance, lack of an active leak detection programme or poor quality of underground assets and unbilled authorised consumption by the utility for operational purposes, for example, water used for firefighting and water provided for free to certain customer groups.[46]

Factors causing this gap include inaccurate billing systems, deficient customer registration, leakage caused by deteriorating infrastructure, poor water pressure management, inaccurate metering, reservoir overflow, unnecessary flushing, insufficient management and illegal connections to the water network. High levels of UFW impact the financial viability of water utilities due to revenue losses and unnecessarily high operating costs. UFW also impacts the capability of utilities to fund necessary maintenance of the system.[47] Reducing UFW provides a range of benefits in addition to increasing revenue. These benefits include the following:

- *The same amount of water for more people*: Reducing UFW will at least postpone the need for additional water resources in cities with a growing population. Up to 30 percent more people can be potentially served by making distribution systems more efficient. Otherwise, growing demand for water in cities will increase the cost of water treatment and pumping, requiring extra plant capacity and more raw water. Also expanding the water distribution network without a leak detection programme is effectively expanding a cycle of inefficiency.

Tokyo reducing UFW

Tokyo aims to achieve the effective use of precious water resources, reduce environmental impacts and improve the efficiency of operations. Tokyo Waterworks actively promotes leakage prevention measures to reduce UFW to just 3 percent. This has been achieved through a process that involves replacing water pipes and improving materials to prevent leakage before it happens and eliminating any existing underground leakage, for example, the utility replaces old service pipes with the introduction of stainless steel service pipes, effectively finding and repairing leakage by scheduling work to detect and repair underground leakage early on district by district and having on standby a 24/7 mobile emergency work system that responds to any aboveground leakage and developing advanced technologies to prevent leakage, for instance, the utility has automatic leak detectors installed between the meter and the sub-main pipe and uses correlative leak detectors to isolate the location of leaks.[49]

- *Lower operational costs*: Not only does UFW mean precious water is lost but so is the energy used to treat and distribute the water. Further energy savings can be obtained by reducing leakage since leakage reduction programmes ensure more stable water pressure through the system, increasing energy efficiency. In addition, reduced pressure and less fluctuation in pressure will extend the life expectancy of pipes and valves.
- *Higher revenues*: High levels of UFW caused by inaccurate metering impact the financial viability of water utilities because of lost revenues – typically up to one quarter of UFW are apparent losses (commercial losses).
- *Safe water quality*: In most cities it is necessary to add chlorine to disinfect water before distribution. But if the distribution system does not work properly, it will result in excessively high chlorine content nearer to pumping stations and low or zero chlorine further out in the distribution system. The flow and age of the distributed water needs to be as uniform as possible over the entire distribution system to make chlorination efficient. If the chlorine degrades the distribution system with high leakage is vulnerable to contamination resulting from vacuums developing during low-pressure situations.[48]

4.2.5   Temporary ordinances and regulations

Water conservation temporary ordinances and regulations restrict certain types of water use during specified times and/or restrict the level of water use to a specified amount. These programmes are usually enacted during times of severe water shortages and cease once the shortage has passed.[50,51] Examples of water-use regulations include restrictions on nonessential water uses, for example, watering lawns, washing cars, filling swimming pools and washing driveways; restrictions on commercial use, for example, carwashes, hotels and other large consumers of water; bans on using water of drinking quality for cooling purposes; and bans on non-recirculating carwashes, laundries and fountains.[52]

Austin Water's temporary restrictions

Austin Water has declared Stage 2 water restrictions as the amount of water in Lakes Travis and Buchanan has fallen below minimum supply level. As part of the restrictions, hosing of gardens and the use of automatic sprinklers may only happen on the customer's watering day. Specifically, hosing can only occur before 10 a.m. and after 7 p.m. on Sunday for residential houses with even addresses and Saturday for houses with odd addresses, while automatic irrigation can only occur before 5 a.m. or after 7 p.m. on Thursday (even addresses) and Wednesday (odd addresses). Hosing by commercial customers can only occur before 10 a.m. or after 7 p.m. on Tuesday (even address) or Friday (odd address) and automatic irrigation systems before 5 a.m. or after 7 p.m. on Tuesday (even address) or Friday (odd address).[53]

CASE 4.7

The city of San Diego's permanent mandatory restrictions

Since 2011, the city of San Diego has permanent mandatory water restrictions in place that apply year-round, irrespective of whether the city is in drought or not. These restrictions are designed to promote water conservation as a permanent way of life in the city. Permanent restrictions include the following: customers must repair or stop all water leaks upon discovery or within 72 hours of notification by the City of San Diego; residents who wash vehicles must implement procedures to conserve water and prevent excessive run-off including washing at a commercial carwash, washing only on a pervious surface or directing water to the lawn; and no new water connections for customers using a single pass-through cooling system.[58]

CASE 4.8

4.2.6   Permanent ordinances and regulations

Water conservation permanent ordinances and regulations include amendments to building codes or ordinances requiring the installation of water-saving devices, for example, low-flow toilets, showerheads and faucets in all newly constructed or renovated homes and offices.[54,55,56] For example, plumbing codes can be used to ensure new homes and offices have maximum water-use standards for plumbing fixtures such as toilets (e.g. must be less than or equal to 1.6 gallons per flush), urinals (e.g. must be less than or equal to 1.0 gallon per flush), faucets and showers (e.g. must be less than or equal to 2.5 gallons per minute at 80 psi or 2.2 gallons per minute at 60 psi).[57]

4.2.7   Source protection

Water utilities are concerned about the quality of their source water. Controlling polluting activities at their source, in contrast to removing them in the drinking water treatment process, reduces human health risks as well as reduces

Source protection in Vienna

Vienna is the first city in the world to constitutionally protect its drinking water. The Vienna Water Charter ensures the city does not expose water to hazards that impact water quality. Vienna's drinking water comes from water springs in the Rax, Schneeberg and Schneealpe mountains and from the Hochschwab mountain massif. To protect the springs the Forestry Office of the City of Vienna maintains source protection forests to ensure the soil remains healthy and able to filter and store rainwater efficiently. In addition, the city works with farmers in the source areas to avoid negative influences on the water sources.[60]

treatment costs. Specifically, from an operations perspective, the better the source water quality, the less money is required for treatment chemicals, equipment and labour. In addition, the less treatment required, the fewer the costs passed on to the water system's customers.[59]

4.2.8   Developing alternative supplies

Developing alternative water supplies, including rainwater harvesting and grey water systems, enables water utilities to reduce the costs of treating potable water for non-potable uses. Rainwater harvesting is the capturing and storing of rainwater for beneficial use. Roof run-off is captured in storage systems and then can be used for non-potable uses including irrigation. This can reduce or eliminate the need for municipal water for landscaping. Grey water means untreated used water that has not come into contact with toilet waste and includes water from showers and laundry machines. Grey water recycling is the reuse of treated grey water for non-potable use including toilet flushing, general washing and irrigation.

4.2.9   Subsidies and rebates

Economic instruments such as subsidies or rebates are used to modify water users' behaviour in a predictable, cost-effective way, that is, reduce wastage and lower water consumption.[61,62,63,64,65] In particular, subsidies (incentives) are commonly used to encourage the uptake of water-saving devices (low-flow toilets, taps and showerheads) or water-efficient appliances (dishwashers and washing machines) as positive incentives are found to be more effective than disincentives in promoting water conservation. In addition, incentives have been found to reduce the gap between the time the incentive is presented and behavioural change as compared to disincentives.[66] In order to accelerate the replacement of old water-using fixtures, water managers also commonly offer rebates to customers who purchase water-efficient devices and appliances.

San Francisco's water efficient equipment retrofits

The San Francisco Public Utilities Commission (SFPUC) offers a pilot assistance programme for nonresidential customers who can significantly reduce their potable water usage through upgrades or replacement of existing on-site indoor water-using equipment. Eligible projects must save at least 200 ccf (149 000 gallons) or more per annum. The SFPUC will provide qualifying projects grant funding of $1.00 per ccf over a 10-year lifespan up to 50 percent of the project's equipment costs, with a maximum amount of $75 000 per project. A single customer may apply for more than one project. The SFPUC will consider two types of equipment retrofits: fixed water-saving retrofit projects consisting of standardised equipment with predictable savings, including commercial laundry retrofits, medical equipment steam sterilisers and cooling tower pH controllers, and custom retrofit projects that consist of unique or site-specific equipment retrofits that result in project-specific water savings. They include any water-saving equipment not listed under the fixed water-saving equipment. Custom retrofit projects are approved on a case-by-case project.[67]

CASE 4.10

4.2.10   Product labelling and retrofits

Water managers can promote water conservation through the use of authoritative schemes such as product labelling as well as managing retrofits of water-saving devices such as taps, showers and toilets in domestic and nondomestic customer's apartments, houses and buildings. In addition, water managers can lead by example in promoting water conservation.

Labelling

The labelling of appliances according to water efficiency is important in reducing water consumption by eliminating unsustainable products from the market; however, this is provided the labelling scheme is clear and comprehensible and identifies both private and public benefits of conserving water. Nonetheless, people are more likely to respond to eco-labels if the environmental benefits match closely personal benefits such as reduced water bills.[68,69,70,71]

Retrofits

Retrofit programmes involve the distribution and installation of replacement devices to physically reduce water use in homes and offices. The most common are toilet retrofits involving customers having their older toilets replaced with newer low-/dual-flush toilets and replacement of showerheads and faucet aerators (devices that when inserted into taps reduce the flow of water).[73,74,75,76] Water-saving devices can be distributed by water managers in numerous ways including door to door with water-saving kits delivered to households, direct installation by trained technicians or plumbers, mass mailing with water-saving

Water Efficiency Labelling and Standards Scheme in Australia

The Water Efficiency Labelling and Standards (WELS) Scheme is Australia's water efficiency labelling scheme that requires certain products to be registered and labelled with their water efficiency, according to the national WELS Act 2005. WELS products include plumbing products (showers, taps, flow controllers), sanitary ware (toilets, urinals) and white goods (washing machines, dishwashers). The WELS Scheme also has minimum water efficiency standards for toilets and washing machines. Retailers cannot sell toilets that have a higher flow rate than 5.5 litres per average flush volume and washing machines that are less than 3 stars for a machine with a 5 kilogram or more capacity or less than 2.5 stars for a machine with less than a 5 kilogram capacity. It is projected that by 2021, the WELS Scheme will help reduce domestic water use by more than 100 000 megalitres per annum, save more than 800 000 megalitres (more water than Sydney Harbour) and reduce greenhouse gas emissions by 400 000 tonnes each year. Over a third of the water savings will come from more efficient showers, around 34 percent from washing machines and 23 percent from more efficient toilets and urinals.[72]

CASE 4.11

South West Water providing free water-saving kits

South West Water in the United Kingdom provides customers with free water-saving kits so they can take control of their water usage. The utility has partnered with save-water.co.uk to offer customers free water-saving devices that can save water and money and reduce their carbon footprint. The kits include a ShowerSave that sets a constant flow rate of 7.5 litres per minute, a shower timer, tap inserts and water-saving tip leaflets.[79]

CASE 4.12

devices posted out, depot pick-up with customers calling in to pick up devices or water-saving device requests where customers request devices for installation.[77] The overall benefits of retrofit programmes are they are relatively inexpensive; easily installed by homeowners or plumbers with little or no disruption to users; effective in reducing water use, waste flows and energy consumption; and permanent.[78]

4.2.11 Service innovation

Consumers are becoming increasingly critical and demanding with respect to the performance of water utilities. Water utilities are now confronted with the challenge of shifting from just being a water supplier to becoming a customer-orientated service provider with a high sustainability profile. As such, water utilities are now deciding on new types of services they wish to include in their future portfolio, for example, including new combinations of water quality for customers.[80]

K-Water's Smart Water Grid

In Korea, good quality water for human and natural use is challenged by climate change extreme weather events and pollution from urbanisation and industrial development. Korea's water agency, K-Water, is responding to uncertainty in water quantity and quality by developing a Smart Water Grid that combines existing water grids with information and communication technologies. The Smart Water Grid enables K-Water to monitor real-time the entire water supply system to ensure adequate quantity and consistent water quality. The Smart Water Grid also comprises a sensor network inside the pipelines that collects and analyses water data including quantity, quality, pressure, leakage and so on. In addition the Smart Water Grid enables customers to receive real-time information about tap water quality over the whole production and transportation process.[81]

CASE 4.13

4.3 Communication and information demand management instruments

Communication and information instruments encourage a water-orientated society. In particular, communication and information tools aim to change behaviour through public awareness campaigns around the need to conserve scarce water resources.

4.3.1 Education and public awareness

Education of the public is crucial in generating an understanding of water scarcity and creating acceptance of the need to implement water conservation programmes.[82]

Education in schools

Water managers can promote water conservation in schools to increase young people's knowledge on the water cycle and encourage the sustainable use of scarce water resources.[83,84,85] To do so, water managers can use a variety of strategies including school presentations, distribution of water conservation information and materials that can be used in the school curriculum.[86]

Public education/awareness

Water managers can use public education to persuade water users to conserve water resources.[88,89] In particular, water managers can influence water users' attitudes and behaviour towards water resources by increasing their knowledge and awareness of environmental problems associated with water scarcity.[90,91,92] There are multiple tools and formats water managers can use to increase environmental

Scottish Water's H2-O education programme

Scottish Water and the Scottish Professional Football League (SPFL) have teamed up to bring a brand new education programme to selected Scottish primary schools. The H2-O programme focuses on the relationship between water, hydration and physical activity. It combines a physical and mental workout for children that emphasises the need to stay hydrated and physically active. Each 90-minute session will be delivered by a professional football coach who is affiliated with the SPFL. The session is divided into two parts with the first part involving an interactive session focusing on the 'Water Cycle' and 'Our Bodies and Water'. The second part involves a physical session with children dribbling a football around cones in different ways depending on what water-related word is shouted out. To deliver the programme Scottish Water and the SPFL Trust are working with three professional football clubs. The programme is free for schools and will reach around 3000 young people across the three clubs' local areas.[87]

CASE 4.14

San Francisco's bold conservation messages

In 2015, the SFPUC launched its multilingual, citywide education campaign that promotes water conservation through behavioural change – with a difference. The SFPUC's public awareness programmes, that feature in or on newspapers, bus shelters, buses and billboards, aim to stand out with creatively crafted messages that read 'Jiggle it' when looking for leaks, 'Make it a quickie' when having a shower and 'Doing it' by replacing old toilets and getting paid for it. These advertisements follow on from their success in 2014 that saw the SFPUC's 2.6 million water users shooting past the target of cutting use by 10 percent and conserved 12 percent. The results have paid off with average use in San Francisco in April 44 gallons a day resulting in San Francisco needing to conserve 8 percent in 2015, compared to other cities in California that must conserve 30 percent.[100]

CASE 4.15

awareness and water conservation including public information (printed literature distributed or available for the general public, public service announcements and advertisements on billboards, public transportation, television commercials, newspaper articles and advertisements, Internet and social media campaigns), public events (customers can receive information on water conservation tips and receive water-saving devices at conservation workshops, expos, fairs, etc. as people frequently make poor choices with regard to environment-friendly products or services due to misinformation or lack of information) and information in water bills. (Water bills should be understandable enabling customers to easily identify volume of usage, rates and charges. Water bills should be informative enabling customers to compare their current bill with previous bills. Finally, water bills should contain water conservation tips to help customers make informed decisions on future water use.)[93,94,95,96,97,98,99]

4.3.2   Competition between water users

Water managers can increase participation rates in water conservation programmes by promoting competition among water users to achieve specific water consumption targets. Examples of competitions include eliciting commitments to water-saving targets and promoting competition through the water bill. Regarding eliciting commitments, water managers can obtain verbal or written commitments from individuals and communities to achieve specific water-saving targets. Competitions can be formed to compare one community with another and offer winners recognition or prizes for their water-saving achievements.[101,102] The water bill can also be used as a tool for competition between water users. For example, water bills can show a household's water consumption compared to the average household in the neighbourhood, city, province or state.[103,104]

As norms can be made 'salient'/prominent by viewing the behaviour of another person or inferring the actions of others, water managers can provide examples of how individuals and communities have successfully conserved water.[105] This enables water users to draw lessons from successful water conservation efforts, helping establish behaviour change towards scarce water resources.[106]

Water managers can 'reference' other 'communities' water savings as models to emulate or mimic. Alternatively, water managers can use water-saving role models such as community leaders or winners of water-saving competitions as reference points for ideal behaviour that can be emulated or mimicked by others.[107,108,109]

4.3.3   Corporate social responsibility

Water managers play a critical role in providing leadership on conservation for several reasons: first, a failure to exemplify the behavioural changes water managers wish to see will undermine any information or persuasion campaigns water managers attempt to engage in at a future date; second, successful internal water conservation programmes send a strong signal to individuals and businesses about what is possible and that water managers are serious about water conservation; and, third, these initiatives allow water managers to learn invaluable lessons first-hand on the difficulties of achieving water conservation goals.[111]

Water Corporation's school water competition

The Water Corporation of Western Australia holds a 'snap a waterwise' photo competition where students during term are invited to take a photo that depicts the importance of water and its impact on everyone's future, with the winning primary and high schools receiving a $1200 camera and kit each. To enter the competition, students share their photos on the water utility's 'Brag About It' page from which the photo with the most votes wins.[110]

CASE 4.16

Sydney Water reducing energy usage and carbon emissions

Sydney Water aims to cap its carbon emission levels at a stable level and keep its non-renewable electricity purchases in 2020 at below 1998 levels. To do so the utility has implemented an energy efficiency programme that has so far saved around 30 gigawatt hours, the equivalent of saving electricity used by 4100 homes a year. This has been achieved by increasing the efficiency of its wastewater treatment plants, investing in energy-efficient buildings and replacing conventional lighting with LED technology at several sites: changing lights at some of the treatment plants has alone saved the utility $130 000 a year. In addition the utility generates 16 percent of its electricity from renewable sources: enough to power 9000 homes each year. Sydney Water recovers biogas to power its wastewater treatment plants, generates hydropower from treated wastewater passing down a long drop shaft and generates solar power at its main office.[112]

CASE 4.17

4.4   Portfolio of demand management tools

In demand management, the use of price alone cannot be relied on to achieve water conservation targets for three reasons: first, the use of pricing alone to achieve water conservation is economically sensitive. In particular, the pricing of water raises the day-to-day costs of households and businesses. This can be inflationary and attract opposition.[113] In addition, water managers need to take note of the price elasticity of water as it indicates the likely revenue impacts from price changes: if demand is elastic, price increases will decrease demand to the extent the water suppliers' total revenues will actually decrease, while inelastic demand will mean revenues from a price increase will outweigh the losses associated with a decrease in demand.[114] Furthermore, the pricing of water is politically sensitive as people are often sceptical on whether price increases for water is really for environmental reasons or just an additional revenue source for the water utilities or government. As such, price is seldom relied on as the main tool in demand management.[115] Second, studies have shown that when pricing is introduced, individuals may modify their behaviour in the short term, but in the long term people revert back to their old habits.[116] Third, it is usually difficult to assess the effectiveness of pricing alone as it is frequently done in conjunction with a variety of other demand management tools (metering, education, regulations and installation of water-saving devices, etc.).[117] Nonetheless, if water managers rely too much on non-price conservation programmes to reduce demand, revenues could decline significantly, impacting the financial sustainability of providing water supplies (the costs being pumping, treating, distributing and operating and maintaining the water infrastructure). As a result, water managers may be forced to increase the price of water substantially following 'successful' non-price conservation programmes as a way to prevent water utilities from suffering unsustainable losses.[118]

Water managers should therefore use a portfolio of demand management tools (regulatory, informational and market-based) to achieve specific water conservation targets.[119,120] However, the portfolio cannot be composed of a random collection of demand management tools; instead each tool must complement and support the overall strategic vision (water conservation target).[121] Furthermore, when deciding on which demand management tool to be incorporated into the portfolio, a cost–benefit analysis should be conducted for each individual tool before its inclusion in the overall demand management strategy as non-price water conservation programmes incur numerous financial costs, for example, advertising, billing inserts, monitoring and enforcement.[122]

Notes

1. GLOBAL WATER PARTNERSHIP. 2012. *Water Demand Management (WDM) – The Mediterranean Experience*. Technical focus paper [Online]. Available: http://www.gwp.org/en/gwp-in-action/News-and-Activities/Global-Water-Partnership-launches-new-publications-at-World-Water-Week-2012/ (accessed 2 June 2016).
2. SAVENIJE, H. & VAN DER ZAAG, P. 2002. Water as an economic good and demand management: paradigms with pitfalls. *Water International*, 27, 98–104.
3. Ibid.
4. GLOBAL WATER PARTNERSHIP. 2012. *Water Demand Management (WDM) – the Mediterranean experience*. Technical focus paper [Online]. Available: http://www.gwp.org/en/gwp-in-action/News-and-Activities/Global-Water-Partnership-launches-new-publications-at-World-Water-Week-2012/ (accessed 2 June 2016).
5. Ibid.
6. Ibid.
7. GIFFORD, R., KORMOS, C. & MCINTYRE, A. 2011. Behavioral dimensions of climate change: drivers, responses, barriers, and interventions. *Wiley Interdisciplinary Reviews: Climate Change*, 2, 801–827.
8. POLICY RESEARCH INSTITUTE. 2005. *Economic Instruments for Water Demand Management in an Integrated Water Resources Management Framework: Synthesis Report*. Ottawa: Policy Research Initiative.
9. SIBLY, H. 2006. Efficient urban water pricing. *Australian Economic Review*, 39, 227–237.
10. POLICY RESEARCH INSTITUTE. 2005. *Economic Instruments for Water Demand Management in an Integrated Water Resources Management Framework: Synthesis Report*. Ottawa: Policy Research Initiative.
11. ROGERS, P., DE SILVA, R. & BHATIA, R. 2002. Water is an economic good: how to use prices to promote equity, efficiency, and sustainability. *Water Policy*, 4, 1–17.
12. OECD. 2010. *Pricing Water Resources and Water and Sanitation Services*. Paris: OECD.
13. CAP-NET. 2008. *Economics in sustainable water management: training manual and facilitators' guide*. Available: http://www.euwi.net/files/Cap_net_EUWI_FWG_GWP_Manual_Economics_of_water_FINAL.pdf (accessed 9 May 2016).
14. SIBLY, H. 2006. Efficient urban water pricing. *Australian Economic Review*, 39, 227–237.
15. POLICY RESEARCH INSTITUTE. 2005. *Economic Instruments for Water Demand Management in an Integrated Water Resources Management Framework: Synthesis Report*. Ottawa: Policy Research Initiative.

16. CAP-NET. 2008. *Economics in sustainable water management: training manual and facilitators' guide.* Available: http://www.euwi.net/files/Cap_net_EUWI_FWG_GWP_Manual_Economics_of_water_FINAL.pdf (accessed 9 May 2016).

17. OLMSTEAD, S. M. & STAVINS, R. N. 2007. *Managing water demand: price vs. non-price conservation programs.* Pioneer Institute White Paper (39), July 2007.

18. POLICY RESEARCH INSTITUTE. 2005. *Economic Instruments for Water Demand Management in an Integrated Water Resources Management Framework: Synthesis Report.* Ottawa: Policy Research Initiative.

19. ROGERS, P., DE SILVA, R. & BHATIA, R. 2002. Water is an economic good: how to use prices to promote equity, efficiency, and sustainability. *Water Policy*, 4, 1–17.

20. Ibid.

21. VAN DER ZAAG, P. & SAVENIJE, H. 2006. *Water as an Economic Good: The Value of Pricing and the Failure of Markets.* Delft: UNESCO-IHE.

22. SIBLY, H. 2006. Efficient urban water pricing. *Australian Economic Review*, 39, 227–237.

23. CAP-NET. 2008. *Economics in sustainable water management: training manual and facilitators' guide.* Available: http://www.euwi.net/files/Cap_net_EUWI_FWG_GWP_Manual_Economics_of_water_FINAL.pdf (accessed 9 May 2016).

24. ROGERS, P., DE SILVA, R. & BHATIA, R. 2002. Water is an economic good: how to use prices to promote equity, efficiency, and sustainability. *Water Policy*, 4, 1–17.

25. EPA. 2006. *Setting small drinking water system rates for a sustainable future* [Online]. Available: https://www.epa.gov/dwcapacity/resources-setting-small-system-water-rates (accessed 2 June 2016).

26. SIBLY, H. 2006. Efficient urban water pricing. *Australian Economic Review*, 39, 227–237.

27. BITHAS, K. 2008. The sustainable residential water use: sustainability, efficiency and social equity. The European experience. *Ecological Economics*, 68, 221–229.

28. Ibid.

29. Ibid.

30. Ibid.

31. Ibid.

32. VAN DER ZAAG, P. & SAVENIJE, H. 2006. *Water as an Economic Good: The Value of Pricing and the Failure of Markets.* Delft: UNESCO-IHE.

33. POLICY RESEARCH INSTITUTE. 2005. *Economic Instruments for Water Demand Management in an Integrated Water Resources Management Framework: Synthesis Report.* Ottawa: Policy Research Initiative.

34. VAN DER ZAAG, P. & SAVENIJE, H. 2006. *Water as an Economic Good: The Value of Pricing and the Failure of Markets.* Delft: UNESCO-IHE.

35. Ibid.

36. POLICY RESEARCH INSTITUTE. 2005. *Economic Instruments for Water Demand Management in an Integrated Water Resources Management Framework: Synthesis Report.* Ottawa: Policy Research Initiative.

37. ECONOMIC AND SOCIAL RESEARCH COUNCIL. 2008. *Behavioural change and water efficiency.* Available: http://webarchive.nationalarchives.gov.uk/20080821115857/http://esrc.ac.uk/ESRCInfoCentre/about/CI/events/esrcseminar/BehaviouralChangeandWaterEfficiency.aspx?ComponentId=25751&SourcePageId=6066 (accessed 9 May 2016).

38. Ibid.

39. Ibid.

40. POLICY RESEARCH INSTITUTE. 2005. *Economic Instruments for Water Demand Management in an Integrated Water Resources Management Framework: Synthesis Report.* Ottawa: Policy Research Initiative.

41. Ibid.
42. CITY OF CALGARY. 2015. *Water efficiency* [Online]. Available: http://www.calgary.ca/UEP/Water/Pages/Water-conservation/Water-efficiency.aspx (accessed 9 May 2016).
43. TOP, H. J. 2010. *Smart grids and smart water metering in the Netherlands.* EC-ICT for water management.
44. IIWIDGET. 2015. *Smart meters, smart water, smart societies* [Online]. Available: http://www.i-widget.eu/images/pdf/iWIDGET-Project-Flyer-low-res-web_Mar2014.pdf (accessed 9 May 2016).
45. LONG BEACH WATER. 2015. *Long Beach Water announces free smart water meter installation program* [Online]. Available: http://www.lbwater.org/sites/default/files/documents/MEDIA%20RELEASE%20Smart%20Meter%20Announcement.pdf (accessed 9 May 2106).
46. PEDERSEN, J. B. & Klee, P. (ED.IN C.) 2013. *Meeting an increasing demand for water by reducing urban water loss: reducing non-revenue water in water distribution.* The Rethink Water network and Danish Water Forum White Paper.
47. Ibid.
48. Ibid.
49. TOKYO METROPOLITAN GOVERNMENT BUREAU OF WATERWORKS. 2015. *Technology for non-revenue water (NWR) reduction, no. 24* [Online]. Available: http://www.metro.tokyo.jp/ENGLISH/ABOUT/TECH/FILES/ENGLISH/2_Infrastructure.pdf (accessed 9 May 2016).
50. MICHELSEN, A. M., MCGUCKIN, J. T. & STUMPF, D. 1999. Nonprice water conservation programs as a demand management tool. *JAWRA Journal of the American Water Resources Association*, 35, 593–602.
51. CANADA WEST FOUNDATION. 2004. *Drop by Drop: Urban Water Conservation Practices in Western Canada.* Western Cities Project Report [Online]. Available: http://cwf.ca/publications-1/drop-by-drop-urban-water-conservation-practices-in-western-canada (accessed 9 May 2106).
52. EPA. 1998. *Water conservation plan guidelines.* Available: https://www3.epa.gov/watersense/pubs/guide.html (accessed 2 June 2016).
53. AUSTIN WATER. 2015. *Watering restrictions* [Online]. Available: http://www.austintexas.gov/department/watering-restrictions (accessed 9 May 2106).
54. MICHELSEN, A. M., MCGUCKIN, J. T. & STUMPF, D. 1999. Nonprice water conservation programs as a demand management tool. *JAWRA Journal of the American Water Resources Association*, 35, 593–602.
55. OECD. 2011. *Greening Household Behaviour: The Role of Public Policy.* Paris: OECD.
56. PENNSYLVANIA STATE UNIVERSITY. 2010. *Water conservation for communities.* Available: http://pubs.cas.psu.edu/FreePubs/PDFs/AGRS113.pdf (accessed 9 May 2016).
57. Ibid.
58. PUBLIC UTILITIES: WATER CITY OF SAN DIEGO. 2015. *Drought alert: mandatory water use restrictions* [Online]. Available: https://www.sandiego.gov/water/conservation/drought/prohibitions (accessed 2 June 2016).
59. WATER RESEARCH FOUNDATION AND EPA. 2010. *Drinking Water Source protection through effective use of TMDL processes.* Available: http://www.waterrf.org/PublicReport Library/4007.pdf (accessed 8 August 2016).
60. CITY OF VIENNA. 2015. *Environment* [Online]. Available: https://www.wien.gv.at/english/environment/protection/reports/pdf/green-04.pdf (accessed 2 June 2016).
61. GLOBAL WATER PARTNERSHIP. 2012. *Water Demand Management (WDM) – the Mediterranean experience.* Technical focus paper [Online]. Available: http://www.gwp.org/en/gwp-in-action/News-and-Activities/Global-Water-Partnership-launches-new-publications-at-World-Water-Week-2012/ (accessed 2 June 2016).

62. POLICY RESEARCH INSTITUTE. 2005. *Economic Instruments for Water Demand Management in an Integrated Water Resources Management Framework: Synthesis Report.* Ottawa: Policy Research Initiative.

63. SAVENIJE, H. & VAN DER ZAAG, P. 2002. Water as an economic good and demand management: paradigms with pitfalls. *Water International,* 27, 98–104.

64. OECD. 2012b. *Environmental Outlook to 2050: The Consequences of Inaction.* Paris: OECD.

65. GLOBAL WATER PARTNERSHIP. 2012. *Water Demand Management (WDM) – the Mediterranean experience.* Technical focus paper [Online]. Available: http://www.gwp.org/en/gwp-in-action/News-and-Activities/Global-Water-Partnership-launches-new-publications-at-World-Water-Week-2012/ (accessed 2 June 2016).

66. POLICY RESEARCH INSTITUTE. 2005. *Economic Instruments for Water Demand Management in an Integrated Water Resources Management Framework: Synthesis Report.* Ottawa: Policy Research Initiative.

67. SAN FRANSCISCO PUBLIC UTILITIES COMMISSION. 2015b. *Grant assistance for water efficient equipment retrofits: New!* [Online]. Available: http://www.sfwater.org/index.aspx?page=512 (accessed 9 May 2106).

68. AMERICAN PSYCHOLOGICAL ASSOCIATION. 2009. *Psychology and global climate change: addressing a multi-faceted phenomenon and set of challenges.* Task Force on the Interface Between Psychology and Global Climate Change [Online]. Available: http://www.apa.org/science/about/publications/climate-change.aspx (accessed 9 May 2016).

69. OECD. 2012b. *Environmental Outlook to 2050: The Consequences of Inaction.* Paris: OECD.

70. OECD. 2008. *Promoting Sustainable Consumption: Good Practices in OECD Countries.* Paris: OECD.

71. OECD. 2011. *Greening Household Behaviour: The Role of Public Policy.* Paris: OECD Publishing.

72. AUSTRALIAN GOVERNMENT. 2015. *Water efficiency labelling and standards (WELS) scheme* [Online]. Available: http://www.waterrating.gov.au/ (accessed 9 May 2016).

73. GEORGIA ENVIRONMENTAL PROTECTION DIVISION WATERSHED PROTECTION BRANCH. 2007. *Water conservation education programs EPD guidance document.* Available: http://www1.gadnr.org/cws/Documents/Conservation_Education.pdf (accessed 9 May 2016).

74. CANADA WEST FOUNDATION. 2004. *Drop by Drop: Urban Water Conservation Practices in Western Canada.* Western Cities Project Report. Calgary: Canada West Foundation [Online]. Available: http://cwf.ca/publications-1/drop-by-drop-urban-water-conservation-practices-in-western-canada (accessed 9 May 2016).

75. MICHELSEN, A. M., MCGUCKIN, J. T. & STUMPF, D. 1999. Nonprice water conservation programs as a demand management tool. *JAWRA Journal of the American Water Resources Association,* 35, 593–602.

76. PENNSYLVANIA STATE UNIVERSITY. 2010. *Water conservation for communities.* Available: http://pubs.cas.psu.edu/FreePubs/PDFs/AGRS113.pdf (accessed 9 May 2106).

77. Ibid.

78. Ibid.

79. SOUTH WEST WATER. 2015. *Free water saving kit* [Online]. Available: http://www.southwestwater.co.uk/freewatersavingkit (accessed 9 May 2106).

80. HEGGER, D. L. T., SPAARGAREN, G., VAN VLIET, B. J. M. & FRIJNS, J. 2011. Consumer-inclusive innovation strategies for the Dutch water supply sector: opportunities for more sustainable products and services. *NJAS – Wageningen Journal of Life Sciences,* 58, 49–56.

81. K-WATER. 2014. *Water for a happier world. K-Water sustainability report.* Available: http://english.kwater.or.kr/web/eng/download/smreport/2014_SMReport.pdf (accessed 9 May 2016).

82. GEORGIA ENVIRONMENTAL PROTECTION DIVISION WATERSHED PROTECTION BRANCH. 2007. *Water conservation education programs EPD guidance document.* Available: http://www1.gadnr.org/cws/Documents/Conservation_Education.pdf (accessed 9 May 2016).

83. KERAMITSOGLOU, K. M. & TSAGARAKIS, K. P. 2011. Raising effective awareness for domestic water saving: evidence from an environmental educational programme in Greece. *Water Policy*, 13, 828–844.

84. OECD. 2012b. *Environmental Outlook to 2050: The Consequences of Inaction.* Paris: OECD.

85. GEORGIA ENVIRONMENTAL PROTECTION DIVISION WATERSHED PROTECTION BRANCH. 2007. *Water conservation education programs EPD guidance document.* Available: http://www1.gadnr.org/cws/Documents/Conservation_Education.pdf (accessed 9 May 2016).

86. MICHELSEN, A. M., MCGUCKIN, J. T. & STUMPF, D. 1999. Nonprice water conservation programs as a demand management tool. *JAWRA Journal of the American Water Resources Association*, 35, 593–602.

87. SCOTTISH WATER. 2015. *What we doing.* Dunfermline, Scotland: Scottish Water.

88. STEG, L. & VLEK, C. 2009. Encouraging pro-environmental behaviour: an integrative review and research agenda. *Journal of Environmental Psychology*, 29, 309–317.

89. VAN ROON, M. 2007. Water localisation and reclamation: steps towards low impact urban design and development. *Journal of Environmental Management*, 83, 437–447.

90. STEG, L. & VLEK, C. 2009. Encouraging pro-environmental behaviour: an integrative review and research agenda. *Journal of Environmental Psychology*, 29, 309–317.

91. NAJJAR, K. & COLLIER, C. R. 2011. Integrated water resources management: bringing it all together. *Water Resources Impact*, 13, 3–8.

92. POLICY RESEARCH INSTITUTE. 2005. *Economic Instruments for Water Demand Management in an Integrated Water Resources Management Framework: Synthesis Report.* Ottawa: Policy Research Initiative.

93. EPA. 1998. *Water conservation plan guidelines.* Available: https://www3.epa.gov/watersense/pubs/guide.html (accessed 2 June 2016).

94. MICHELSEN, A. M., MCGUCKIN, J. T. & STUMPF, D. 1999. Nonprice water conservation programs as a demand management tool. *JAWRA Journal of the American Water Resources Association*, 35, 593–602.

95. PENNSYLVANIA STATE UNIVERSITY. 2010. *Water conservation for communities.* Available: http://pubs.cas.psu.edu/FreePubs/PDFs/AGRS113.pdf (accessed 9 May 2016).

96. THE STATE OF ISRAEL MINISTRY OF NATIONAL INFRASTRUCTURES PLANNING DEPARTMENT WATER AUTHORITY. 2011. *The State of Israel: National Water Efficiency Report.* Available: http://www.water.gov.il/Hebrew/ProfessionalInfoAnd Data/2012/04-The-State-of-Israel-National-Water-Efficiency-Report.pdf (accessed 9 May 2016).

97. GEORGIA ENVIRONMENTAL PROTECTION DIVISION WATERSHED PROTECTION BRANCH. 2007. *Water conservation education programs EPD guidance document.* Available: http://www1.gadnr.org/cws/Documents/Conservation_Education.pdf (accessed 9 May 2016).

98. BIO INTELLIGENCE SERVICE. 2012. *Policies to encourage sustainable consumption,* Final report prepared for European Commission (DG ENV). Available: http://ec.europa.eu/environment/eussd/pdf/report_22082012.pdf (accessed 9 May 2016).

99. KERAMITSOGLOU, K. M. & TSAGARAKIS, K. P. 2011. Raising effective awareness for domestic water saving: evidence from an environmental educational programme in Greece. *Water Policy*, 13, 828–844.

100. SAN FRANSCISCO PUBLIC UTILITIES COMMISSION. 2015a. *Conservation* [Online]. Available: http://www.sfwater.org/index.aspx?page=136 (accessed 9 May 2106).

101. POLICY RESEARCH INSTITUTE. 2005. *Economic Instruments for Water Demand Management in an Integrated Water Resources Management Framework: Synthesis Report*. Ottawa: Policy Research Initiative.

102. PATCHEN, M. 2010. What shapes public reactions to climate change? Overview of research and policy implications. *Analyses of Social Issues and Public Policy*, 10, 47–68.

103. GEORGIA ENVIRONMENTAL PROTECTION DIVISION WATERSHED PROTECTION BRANCH. 2007. *Water conservation education programs EPD guidance document*. Available: http://www1.gadnr.org/cws/Documents/Conservation_Education.pdf (accessed 9 May 2016).

104. PATCHEN, M. 2010. What shapes public reactions to climate change? Overview of research and policy implications. *Analyses of Social Issues and Public Policy*, 10, 47–68.

105. GEORGIA ENVIRONMENTAL PROTECTION DIVISION WATERSHED PROTECTION BRANCH. 2007. *Water conservation education programs EPD guidance document*. Available: http://www1.gadnr.org/cws/Documents/Conservation_Education.pdf (accessed 9 May 2016).

106. FRANTZ, C. M. & MAYER, F. S. 2009. The emergency of climate change: why are we failing to take action? *Analyses of Social Issues and Public Policy*, 9, 205–222.

107. STEG, L. & VLEK, C. 2009. Encouraging pro-environmental behaviour: an integrative review and research agenda. *Journal of Environmental Psychology*, 29, 309–317.

108. OECD. 2012a. *Behavioural economics and environmental policy design*. Available: http://www.oecd.org/env/consumption-innovation/Behavioural%20Economics%20 and%20Environmental%20Policy%20Design.pdf (accessed 9 May 2016).

109. CORREIA, R. & ROSETA-PALMA, C. 2012. *Behavioural economics in water management*. Available: http://www.isee2012.org/anais/pdf/742.pdf (accessed 9 May 2016).

110. WATER CORPORATION. 2015a. *Competition for schools* [Online]. Available: http://www.watercorporation.com.au/home/teachers/grants-and-competitions/competitions-for-schools (accessed 9 May 2016).

111. JACKSON, T. 2005. *Motivating sustainable consumption: a review of evidence on consumer behaviour and behavioural change: a report to the Sustainable Development Research Network*. Centre for Environmental Strategy, University of Surrey.

112. SYDNEY WATER. 2015a. *Energy management and climate change* [Online]. Available: https://www.sydneywater.com.au/SW/water-the-environment/what-we-re-doing/ energy-management/index.htm (accessed 9 May 2016).

113. CAP-NET. 2008. *Economics in sustainable water management: training manual and facilitators' guide*. Available: http://www.euwi.net/files/Cap_net_EUWI_FWG_GWP_ Manual_Economics_of_water_FINAL.pdf (accessed 9 May 2016).

114. OLMSTEAD, S. M. & STAVINS, R. N. 2007. *Managing water demand: price vs. non-price conservation programs*. Pioneer Institute White Paper (39).

115. POLICY RESEARCH INSTITUTE. 2005. *Economic Instruments for Water Demand Management in an Integrated Water Resources Management Framework: Synthesis Report*. Ottawa: Policy Research Initiative.

116. Ibid.

117. Ibid.

118. OLMSTEAD, S. M. & STAVINS, R. N. 2007. *Managing water demand: price vs. non-price conservation programs.* Pioneer Institute White Paper (39).

119. CAP-NET. 2008. *Economics in sustainable water management: training manual and facilitators' guide.* Available: http://www.euwi.net/files/Cap_net_EUWI_FWG_GWP_Manual_Economics_of_water_FINAL.pdf (accessed 9 May 2016).

120. OECD. 2012b. *Environmental Outlook to 2050: The Consequences of Inaction.* Paris: OECD.

121. POLICY RESEARCH INSTITUTE. 2005. *Economic Instruments for Water Demand Management in an Integrated Water Resources Management Framework: Synthesis Report.* Ottawa: Policy Research Initiative.

122. OLMSTEAD, S. M. & STAVINS, R. N. 2007. *Managing water demand: price vs. non-price conservation programs.* Pioneer Institute White Paper (39).

5 Transitions

Introduction

To achieve urban water security, water utilities will need to transition towards the sustainable use of water that balances demand with supply. However, what exactly is a transition and how can a transition be implemented to achieve a desired outcome? This chapter first defines what a transition is, what types of transitions there are, how they occur over multiple dimensions and the various drivers and forces of transitions before finally discussing how transitions can be managed.

5.1 What is a transition?

A transition is a well-planned shift from one sociotechnical system to another, over a period of one to two generations,[1] where a sociotechnical system is a stable configuration of human and non-human elements including technology, regulations, markets, user practices and cultural meanings, infrastructure, maintenance and supply networks (Table 5.1).[2] In daily life, sociotechnical systems serve societal functions such as water, energy and transportation systems.[3,4,5,6,7] The term 'sociotechnical' is used to stress the influence of technology on society and the influence of society on technology.[8] Overall, a transition is a gradual, yet continuous, process in which the structural character of society, in a sociotechnical system, fundamentally changes over a long period of time.[9] This structural change

Urban Water Security, First Edition. Robert C. Brears.
© 2017 John Wiley & Sons, Ltd. Published 2017 by John Wiley & Sons, Ltd.

Table 5.1 Components of a sociotechnical system

Sociotechnical regime components	
Users	Domestic/non-domestic users and different sectors of the economy
Societal groups	Advocacy groups, public authorities, research institutes
Public authorities	Public utilities, local governments, national governments
Research	Universities, technical institutions, R&D labs
Production	Firms, engineers, designers
Finance	Banks, venture capital/investment firms
Supply chain	Material suppliers, component/machine suppliers

GEELS, F. W. 2005. Processes and patterns in transitions and system innovations: refining the co-evolutionary multi-level perspective. *Technological Forecasting and Social Change*, 72, 682

occurs through a combination of behavioural, cultural, ecological, economic, institutional and technological developments that positively reinforce one another for change to occur.[10,11,12,13,14,15,16,17,18,19,20,21,22,23]

In transitions, the role of institutions is to first create a futuristic vision of what the new sociotechnical system looks like and second coordinate the appropriate resources (economic, financial, knowledge, etc.) to achieve that vision.[24,25] In particular, the role of institutions in transitions is to ensure the necessary resources required for transitions are available (including factor endowments, capabilities and knowledge) and manage the coordinated deployment of these resources.[26] The prerequisite for a transition to be successful is that all of these developments positively reinforce one another for change to occur.[27]

5.1.1 What types of transitions are there?

There are four types of transitions: endogenous renewals (regime actors make a conscious and planned transition in response to perceived pressures using regime internal resources), reorientation of trajectories (results from a shock (internal or external) on the incumbent regime, followed by a response from regime actors using internal resources), emergent transformations (arises from uncoordinated pressures outside of the regime) and purposeful transformations that are deliberately intended and pursued from the outset to reflect explicit social expectations or interests.[28]

5.1.2 Transitions occur over multiple dimensions

A transition from one sociotechnical system to another is a process that involves multiple dimensions and levels, each reinforcing one another over the transition period.[29] Specifically, transitions are multi-actor as they involve a wide range of actors (firms, consumers, non-governmental organisations, knowledge producers and governments) and multifactor as they involve the interplay of many factors that influence one another in the process of change (technical, regulatory, societal and behavioural). Transitions are also multilevel (multilevel perspective) in that change occurs in the system over multiple levels: The macro (landscape), meso (regime) and the micro (individuals) with the relationship between the three levels being a nested hierarchy with the regime embedded between the landscape (macro level) and the niche (micro level).[30,31]

The macro level, or landscape, is the surroundings or the environment in which a system operates and is beyond the direct influence of the meso and micro level.[32,33] The term 'landscape' is frequently used to describe the macro level because it is 'hard' and indicates the material aspects of society.[34] At the macro level, the landscape is determined by exogenous changes in the natural environment and culture. The macro level also contains the institution's visions of how the landscape should look in the future.[35] Sociotechnical landscapes are relatively static as factors do not change or change only slowly. Nonetheless, landscapes can change with rare, rapid external shocks such as large-scale environmental, economic or political shocks. As such, changes at the macro level are generally slow, usually taking place over decades.[36,37]

The meso level comprises the sociotechnical system's regime, which is described as a constellation of cultures, structures and practices of the system's social users (individual users, societal groups, public authorities, research networks, financial institutions, etc.).[38,39] While each of the system's social users is autonomous, they are at the same time interdependent and interact with one another. This interdependence occurs because the activities of the system's social users are coordinated with one another in the running of the sociotechnical system.[40] Therefore, regimes are stable and durable.[41] If a transition is to be successful, it must alter the core beliefs and values of the regime's social users. However, this is difficult as it involves unlearning what has been ingrained such as assumptions, heuristics, norms and beliefs that have been established within individuals and society.[42]

At the micro level, niches comprise innovations (new ideas, alternative practices and innovative technologies) that deviate from the status quo.[43,44,45,46] These innovations occur in many dimensions, for example, technology, user preferences, regulations, symbolic meanings, infrastructural and production systems.[47] However, because innovations are alternatives to the mainstream, it is difficult for them to be accepted on their own standing by wider mainstream audiences.[48] As such, niches at the micro level provide social spaces for new innovations to develop without being exposed to outside pressures and influences.[49,50] In addition, niches enable innovations to gain social networks that support innovations (user–producer relationships).[51] If these niches are successful and the alternative technology or practice is robust, it will branch out and attract mainstream audiences, with the aim of eventually becoming social norms.[52,53]

5.1.3   The transition process

The transition process is non-linear. In particular, transitions follow an 's' curve shape. In the 's' curve, there are four phases of a transition, which are listed as *A*, *B*, *C* and *D* in Figure 5.1:

- *A: Pre-development phase* – The regime (meso level) is often the inhibiting factor to change by maintaining existing social norms and relying on improving existing technologies. Nonetheless, despite very little visible change on the societal level, there are numerous amounts of experimentation taking place.
- *B: Take-off phase* – Developments start to take place at the micro and macro levels with innovations at the micro level being reinforced by changes at the macro level and vice versa. The result is the state of the system beginning to shift.
- *C: Acceleration phase* – Visible structural change occurs rapidly through the accumulation of cultural, economic, ecological/environmental and institutional changes that reinforce one another. This structural change results in the formation of niche regimes which are constellations of cultures, structures and practices that provide a competitive alternative to that of the regime. At the acceleration stage, niche regimes replace the current regime when a critical mass of actors changes their behaviour (the 'tipping point').
- *D: Stabilisation phase* – Social changes decrease and a new equilibrium is reached, which in itself can contain the seeds of change for another transition.[54,55,56,57]

5.1.4   Multilevel drivers of transitions

Before a transition can occur, there first needs to be a misfit or 'gap' between individuals and society's deeply held values and the current conditions they face.[58,59,60]

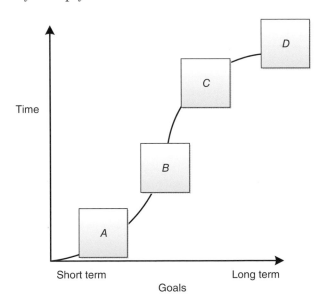

Figure 5.1 The transition 's' curve.

In the multilevel perspective of transitions, institutions create gaps at multiple levels. At the macro level, institutions can create tension with the meso level (regime) by creating a gap between the new strategic vision of the future and the current regime's outdated practices, while at the micro level, institutions place pressure on the meso level through innovations that attempt to create a gap by creating a new alternative regime (niche regime) to the current, outdated regime.[61,62]

Transitions can also be triggered by changes in the external environment of the system leading to it being inefficient, ineffective or inadequate in fulfilling its societal function. As such, external triggers can throw the previous practices of the regime into discredit, creating a gap between the regime's values and the current conditions the system faces.[63,64] These external drivers are social, technological, economic, environmental and political (STEEP) (listed in Table 5.2).

Table 5.2 STEEP drivers of transitions

Driver	Examples
Social	Population growth, urbanisation, demand for cleaner environments, demographic changes
Technological	New technologies/technological innovations that help or hinder efforts of society
Economic	Economic growth, economic shocks, infrastructure growth, economic competition
Environmental	Climate change, environmental degradation, change in land cover and land use, natural disasters
Political	International commitments (Rio 1992, Agenda 21), environmental laws and regulations, transboundary nature of environmental problems

AMERICAN PSYCHOLOGICAL ASSOCIATION, TASK FORCE ON THE INTERFACE BETWEEN PSYCHOLOGY AND GLOBAL CLIMATE CHANGE. 2009. Psychology and Global Climate Change: Addressing a Multi-faceted Phenomenon and Set of Challenges. Washington, DC: American Psychological Association [Online]. Available: http://www.apa.org/science/about/publications/climate-change.aspx (accessed 10 May 2016). OECD. 2012. OECD Environmental Outlook to 2050: The Consequences of Inaction. Paris: OECD. EUROPEAN ENVIRONMENTAL AGENCY. 2013. *Adaptation in Europe: addressing risks and opportunities from climate change in the context of socio-economic developments.* Available: http://www.eea.europa.eu/publications/adaptation-in-europe (accessed 10 May 2016). KOTLER, P. & LEE, N. 2008. Social Marketing: Influencing Behaviors for Good. Los Angeles: Sage Publications. SEYFANG, G. & SMITH, A. 2007. Grassroots innovations for sustainable development: towards a new research and policy agenda. *Environmental Politics*, 16, 584–603. SMITH, A., STIRLING, A. & BERKHOUT, F. 2005. The governance of sustainable socio-technical transitions. *Research Policy*, 34, 1491–1510. JACKSON, T. 2005. *Motivating sustainable consumption: a review of evidence on consumer behaviour and behavioural change: a report to the Sustainable Development Research Network.* University of Surrey, Guildford. VOORA, V. A. & VENEMA, H. D. 2008. The Natural Capital Approach: A Concept Paper. Winnipeg: International Institute for Sustainable Development. ENGEL, K., JOKIEL, D., KRALJEVIC, A., GEIGER, M., & SMITH, K. 2011. Big Cities. Big Water. Big Challenges: Water in an Urbanizing World. Berlin: WWF Germany. Available: www.wwf.se/source.php?id=139089 (accessed 2 June 2016). VOORA, V. A. & VENEMA, H. D. 2008. The Natural Capital Approach: A Concept Paper. Winnipeg: International Institute for Sustainable Development. PELLING, M. 2010. Adaptation to Climate Change: From Resilience to Transformation. London: Taylor & Francis. SCHULTZ, P. 2011. Conservation means behavior. *Conservation Biology*, 25, 1080–1083. DÜRRSCHMIDT, J. 2002. Multiple agoras: local and regional environmental policies between globalization and European pathways of transformation. *Innovation: The European Journal of Social Science Research*, 15, 193–209. KEINER, M. & KIM, A. 2007. Transnational city networks for sustainability. *European Planning Studies*, 15, 1369–1395

5.1.5   Forces in transitions

For a transition to occur – closing of these gaps – there needs to be forces that provide direction for the tensions and pressures placed on the meso level.[65] There are two types of forces that direct transitions: supportive and formative. Supportive forces are top-down (macro level) forces that create tension with the regime by standardising practices or routines through standards and directives. This ensures that practices or routines enjoy universal status by enabling the provision of services (subsidies, capital, investments, etc.) to empower and scale up innovations at the micro level so they become alternatives to the current regime.[66] Formative forces are bottom-up (micro level) forces that create pressure on the regime through innovations or groups of actors adopting innovative practices, routines, services or technology. These innovations have the potential to scale up and challenge the existing regime. Formative forces can emerge naturally or be artificially created by institutions.[67]

5.2   Operationalisation of transitions

In transitions, the application of supportive forces at the macro level can take the form of alternative visions of the future which frame problems and motivate actors to solve them.[68] Even when end points are highly contested, or only partially understood, ideas about what might or should be are essential to envisioning the possibility and motivating change. Alternative visions of the future play a number of functions:

- *Mapping a possibility space*: Visions identify plausible alternatives and means of achieving them.
- *Acting as problem-defining tools*: Visions point to technical, institutional and behavioural problems that need to be resolved.
- *Enabling target setting and monitoring of progress*: Visions stabilise technical and other innovative activity by serving as a common reference point for actors collaborating on its realisation.
- *Acting as a metaphor for building actor networks*: Visions specify relevant actors and act as symbols that bind together communities of interest and of practice.
- *Providing a narrative for focusing capital and other resources*: Visions become an emblem for the marshalling of resources from outside the regime.[69]

At the micro level, the application of formative forces can take the form of diffusion, which is a process where ideas, norms and innovations are communicated over time among members of a social system.[70,71,72] The aim of diffusion is to initiate social change, in particular, the altering of society's norms and values[73]: norms provide a range of tolerable behaviour and serve as guidelines or standards for the behaviour of members of a social system.[74] As such, the norms of a system tell individuals what behaviour they are expected to perform,[75] while values are defined as belief systems, problem definitions and guiding principles that enable individuals to select those elements of reality to which attention should be given, a rationality where there is a structure to evaluate what is logic/illogic, a morality on what is morally right and wrong and prescriptivity which prescribes, implicitly or explicitly, the desirability/undesirability of possible courses of action.[76] Overall, values are important and enduring beliefs or ideas that are shared by members of a particular community. As values underpin a person's decisions and actions, long-term behaviour change can be achieved through a change of values over time.[77,78]

5.2.1   Approaches in decision-making

Diffusion is not an automatic process; instead, it requires the active promotion by institutions.[79] In order to plan and manage the process of diffusion, institutions need to understand the mechanisms through which ideas and practices are spread.[80] However, before these mechanisms can be implemented, institutions need to understand that there are two approaches as to how individuals make decisions: rational and constructivist.[81,82] In the rational approach, individuals are assumed to be rational and goal oriented. Through their actions individuals aim to maximise their utilities by weighing up the costs/benefits of different options before 'actioning' a decision (the logic of consequentialism).[83] In the rational approach, actor's preferences over means/actions and strategies may change but not the preferences over ends and outcomes.[84] In the constructivist approach, individuals are not always rational in their decision-making processes. Instead, their decisions are guided by beliefs, emotions, judgments and morals which themselves are guided by collectively shared understandings of what is considered proper and socially acceptable behaviour (logic of appropriateness).[85,86,87,88] Actors can follow the logic of appropriateness in two ways: first, actors can behave appropriately by learning how to act in accordance with expectations irrespective of whether they agree or not with those expectations. The key is when an actor knows what is socially acceptable in a given setting, group or community; by this, actors are conscious role players. Second, actors accept community norms as the right thing to do. In this situation, actors adopt the interests or even adopt the identity of the community in which they wish to be connected with.[89,90]

5.2.2   Diffusion strategies

Using the rationalist/constructivist approach, institutions can use two types of diffusion strategies to influence the norms and values of society: antecedent and consequential strategies. Antecedent strategies attempt to influence the determinants of target behaviour through activities such as increasing the actor's knowledge or awareness of problems through information campaigns, behavioural commitments and prompting; the assumption being these strategies can influence the determinants of behaviour before its performance. Consequential strategies (feedback, rewards and punishments) are assumed to influence the determinants of target behaviour *after* the performance of the behaviour. In particular, the strategy assumes that feedback, both positive and negative, of the consequences of that behaviour influences the likelihood of that behaviour being performed again in the future.[91]

5.3 Diffusion mechanisms

In the process of diffusion, there are two types of diffusion mechanisms that can induce social change: direct and indirect. In direct diffusion, institutions actively promote ideas, norms and values, while indirect diffusion involves actors, independently, emulating best practices and solutions that serve their needs.[92]

5.3.1 Direct diffusion mechanisms

In direct diffusion, institutions can use manipulation of utility calculations, legal or physical coercion, socialisation and persuasion to induce social change.

Manipulation of utility calculations

Institutions can induce behavioural and social change through the changing of actors' utility functions. The most common tool institutions use to induce social change is market-based instruments such as pricing and subsidies.[93,94]

Legal or physical coercion

Institutions can influence human behaviour through laws, directives and regulations.[95] In particular, these tools are essentially command tools that obligate compliance by targeted audiences.[96]

Socialisation

Institutions promote rules, norms, ideas and practices through the provision of authoritative models. Actors will then seek to meet social expectations of a given community the actor identifies with by internalising them into their domestic structures.[97,98,99]

Persuasion

In persuasion, institutions persuade individuals and society to achieve goals deemed to be in the public interest.[100] Specifically, institutions achieve goals by promoting ideas as legitimate or true through reasoning.[101]

5.3.2 Indirect diffusion mechanisms

The process of diffusion can also occur through indirect mechanisms where actors independently emulate best practices and institutional solutions that serve their needs.[102] In indirect diffusion, there are several ways in which institutions may affect domestic change without the active promotion of ideas: competition, lesson drawing and emulation.

Competition

Individuals independently adjust their behaviour towards 'best practices' which in turn promotes competition between individuals. In particular, individuals borrow ideas from one another to improve their performance (functional emulation) in comparison with others.[103]

Lesson drawing

Actors adopt particular aspects of institutional solutions that are most appropriate for the local context.[104]

Emulation and mimicry

Individuals emulate others in order to be seen as a legitimate member of a community, while mimicry involves the automatic downloading of 'institutional software' without modification, simply because the individual is simply doing 'what everyone else does'.[105]

5.3.3 The diffusion process

The success of an innovation in diffusion depends on its:

1 *Relative advantage*: The degree that an innovation is perceived to be better than the idea it supersedes
2 *Compatibility*: The degree to which an innovation is perceived as being consistent with existing values, past experiences and the needs of potential adopters
3 *Complexibility*: The degree that an innovation is perceived as difficult to understand and use
4 *Trialability*: The degree that an innovation may be experimented within a limited basis
5 *Observability*: The degree to which results of an innovation are visible to others.[106]

5.3.4 Lock-in and barriers to diffusion

Sociotechnical systems are stable configurations of technologies, regulations, standards, institutional tools, lifestyles, investments in machines, infrastructures and competencies, all of which interact and stabilise one another in multiple ways. As such, existing sociotechnical systems 'lock-in' existing trajectories (path dependencies), where current outcomes depend on previous outcomes, and lock-out alternative trajectories that could be more appropriate in meeting current conditions a sociotechnical system faces.[107,108,109,110,111,112,113,114] This lock-in occurs because the existing regime in a sociotechnical system is entrenched in many ways (institutionally, organisationally, economically, culturally, etc.).[115]

In the process of diffusion, there are barriers at work that slow down the process of diffusion. In particular, there are numerous structural (external) and social and psychological (internal) barriers that slow down the process of diffusion.[116,117,118,119,120] Without detailed knowledge of the barriers present, it is unlikely that strategies to modify the norms and values of society will work.[121] As such, it is important that barriers are identified as best as possible as these enable more targeted diffusion mechanisms to be enacted to initiate social change.[122,123,124]

5.4   Transition management

Transitions are complex changes involving multifactors, actors and levels over long periods of time; therefore, they are often beyond the scope of an institution's control, for example, institutions cannot directly control behavioural change in individuals and society. However, the direction, speed and fostering of a transition, and therefore the odds that a transition will occur, can all be influenced by institutions adapting, influencing and monitoring the process.[125,126,127] The guiding philosophy of transitions is that they are goal-oriented, rather than command and control, processes.[128] Transitions also need coordinating as successful activity at one level is not enough to generate transitions; instead, it must be instigated at the multilevel.[129] Because transitions occur over multiple dimensions, there is a high level of uncertainty and unpredictability of whether a transition will succeed or not.[130] To ensure transitions have a better probability of succeeding, transitions can be combined into a management strategy that seeks fundamental changes in the behaviour, cultures, structures and practices of a society.[131,132] This strategy is known as transition management.

5.4.1   Transition management levels

Transition management is a strategy that guides when and how transitions can be initiated, facilitated and influenced.[133,134] In particular, transition management is based on coordinating multi-actor processes at different levels.[135] Transition management has three different types of governance levels in the context of societal transitions: strategic, tactical and operational. The strategic level involves all activities and developments that aim to change, collectively, the norms and values of society as a whole.[136] At this level the problem is defined and visions of alternative futures are created. For these visions to be achieved in transitions, they need

Table 5.3 Transition management actions at multiple levels

Transition management types	Focus	Problem scope	Timescale
Strategic	Culture	Societal	Long term (30+ years)
Tactical	Structure	Regimes	Midterm (5–15 years)
Operational	Process	Project	Short term (0–5 years)

LOORBACH, D. 2010. Transition management for sustainable development: a prescriptive, complexity-based governance framework. *Governance*, 23, 161–183

to offer an inspiring image of a future state; one that contains a behavioural component that is both appealing and imaginative from which it is translated into institutional, economic, ecological and sociocultural aspects associated with the final image.[137] Activities at the tactical level are related to the removal of barriers in the regime that inhibit the achievement of a particular goal. At the tactical level, institutions attempt to modify and remove barriers through enacting changes in rules, regulations, institutions, infrastructure and routines.[138,139] Activities at the operational level relate to short-term experiments carried out in the context of innovative projects and programmes.[140] This includes all societal, technological, institutional and behavioural experiments that introduce or operationalise new structures, cultures and routines. Action at this level is driven by norm entrepreneurs who seek to broaden, deepen or scale up existing and planned initiatives and actions via experiments.[141] Experiments at the operational level need to fit into the context of the strategic vision. In addition, experiments also need to be iconic and involve high levels of risk because success means large innovative contributions towards the meeting of the transition vision.[142] To reduce the risk of failure, institutions can create a portfolio of experiments that can compete with one another, complement one another or investigate various pathways of achieving the strategic vision. Finally, transition management recognises that actions at the strategic, tactical and operational level have differing time horizons as specified in Table 5.3.

5.4.2   Coordination of activities across the levels

An important aspect of transition management is that activities at the strategic, tactical and operational level run in parallel and are coordinated in order to scale up micro-level innovations. If there is too little interaction between the levels, alternative norms and values will become isolated and fail to take off (become mainstream).[143] To ensure activities at all levels run in parallel, instead of sequentially, institutions can utilise the transition management cycle which provides a guideline for the logical order of transitions to achieve long-term visions.[144]

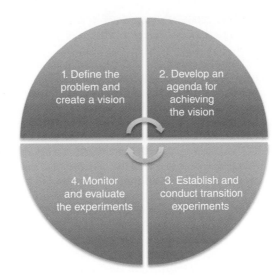

Figure 5.2 Transition management cycle.

5.4.3   Transition management cycle

The transition management cycle provides a blueprint of how institutions can plan, implement, monitor and evaluate transitions. The transition management cycle consists of four parts, summarised in Figure 5.2:

1 *Define the problem and create a vision* of an alternative landscape or future.
2 *Develop an agenda for achieving that vision* by selecting a target behaviour that has the largest impact on achieving the vision and identify the barriers to achieving it.
3 *Establish and conduct transition experiments* that provide new solutions and instruments to achieving the vision.
4 *Monitor and evaluate the experiments* regularly and learn before adjusting, where necessary, the vision and agenda. If experiments are successful, as defined by its contribution towards achieving the transition vision, they can be repeated in different contexts (broadening) and scaled up (deepening) from the micro to macro level.[145,146,147,148]

By following the transition management cycle, institutions can learn from experiments and constantly assess and periodically adjust policies for achieving the vision.[149,150,151]

Notes

1. GEELS, F. W. & SCHOT, J. 2007. Typology of sociotechnical transition pathways. *Research Policy*, 36, 399–417.

2. GEELS, F. W. 2005. Processes and patterns in transitions and system innovations: refining the co-evolutionary multi-level perspective. *Technological Forecasting and Social Change*, 72, 682.

3. Ibid.

4. SOFOULIS, Z. 2005. Big water, everyday water: a sociotechnical perspective. *Continuum: Journal of Media & Cultural Studies*, 19, 445–463.

5. SMITH, A., STIRLING, A. & BERKHOUT, F. 2005. The governance of sustainable socio-technical transitions. *Research Policy*, 34, 1491–1510.

6. ELZEN, B. & WIECZOREK, A. 2005. Transitions towards sustainability through system innovation. *Technological Forecasting and Social Change*, 72, 651–661.

7. SEYFANG, G. & SMITH, A. 2007. Grassroots innovations for sustainable development: towards a new research and policy agenda. *Environmental Politics*, 16, 584–603.

8. Ibid.

9. ROTMANS, J., KEMP, R. & VAN ASSELT, M. 2001. More evolution than revolution: transition management in public policy. *Foresight*, 3, 15–31.

10. VAN DER BRUGGE, R. & ROTMANS, J. 2007. Towards transition management of European water resources. *Integrated Assessment of Water Resources and Global Change: A North-South Analysis*, 21, 249–267.

11. ROTMANS, J., KEMP, R. & VAN ASSELT, M. 2001. More evolution than revolution: transition management in public policy. *Foresight*, 3, 15–31.

12. VAN DER BRUGGE, R., ROTMANS, J. & LOORBACH, D. 2005. The transition in Dutch water management. *Regional Environmental Change*, 5, 164–176.

13. Ibid.

14. GEELS, F. W. 2005. Processes and patterns in transitions and system innovations: refining the co-evolutionary multi-level perspective. *Technological Forecasting and Social Change*, 72, 682.

15. SEYFANG, G. & SMITH, A. 2007. Grassroots innovations for sustainable development: towards a new research and policy agenda. *Environmental Politics*, 16, 584–603.

16. GEELS, F. W. & SCHOT, J. 2007. Typology of sociotechnical transition pathways. *Research Policy*, 36, 399–417.

17. ELZEN, B. & WIECZOREK, A. 2005. Transitions towards sustainability through system innovation. *Technological Forecasting and Social Change*, 72, 651–661.

18. Ibid.

19. GEELS, F. W. & SCHOT, J. 2007. Typology of sociotechnical transition pathways. *Research Policy*, 36, 399–417.

20. ELZEN, B. & WIECZOREK, A. 2005. Transitions towards sustainability through system innovation. *Technological Forecasting and Social Change*, 72, 651–661.

21. Ibid.

22. GEELS, F. W. 2005. Processes and patterns in transitions and system innovations: refining the co-evolutionary multi-level perspective. *Technological Forecasting and Social Change*, 72, 682.

23. ROTMANS, J., KEMP, R. & VAN ASSELT, M. 2001. More evolution than revolution: transition management in public policy. *Foresight*, 3, 15–31.

24. BARBIER, E. 2011. The policy challenges for green economy and sustainable economic development. *Natural Resources Forum*, 35, 233–245.

25. GEELS, F. W. & SCHOT, J. 2007. Typology of sociotechnical transition pathways. *Research Policy*, 36, 399–417.

26. Ibid.

27. VAN DER BRUGGE, R., ROTMANS, J. & LOORBACH, D. 2005. The transition in Dutch water management. *Regional Environmental Change*, 5, 164–176.

28. GEELS, F. W. & SCHOT, J. 2007. Typology of sociotechnical transition pathways. *Research Policy*, 36, 399–417.
29. GEELS, F. W. 2005. Processes and patterns in transitions and system innovations: refining the co-evolutionary multi-level perspective. *Technological Forecasting and Social Change*, 72, 682.
30. Ibid.
31. ELZEN, B. & WIECZOREK, A. 2005. Transitions towards sustainability through system innovation. *Technological Forecasting and Social Change*, 72, 651–661.
32. FRANTZESKAKI, N. & DE HAAN, H. 2009. Transitions: two steps from theory to policy. *Futures*, 41, 593–606.
33. GEELS, F. W. & SCHOT, J. 2007. Typology of sociotechnical transition pathways. *Research Policy*, 36, 399–417.
34. GEELS, F. W. 2005. Processes and patterns in transitions and system innovations: refining the co-evolutionary multi-level perspective. *Technological Forecasting and Social Change*, 72, 682.
35. MCKENZIE-MOHR, D. 2000. New ways to promote proenvironmental behavior: promoting sustainable behavior: an introduction to community-based social marketing. *Journal of Social Issues*, 56, 543–554.
36. GEELS, F. W. & SCHOT, J. 2007. Typology of sociotechnical transition pathways. *Research Policy*, 36, 399–417.
37. FRANTZESKAKI, N. & DE HAAN, H. 2009. Transitions: two steps from theory to policy. *Futures*, 41, 593–606.
38. GEELS, F. W. 2005. Processes and patterns in transitions and system innovations: refining the co-evolutionary multi-level perspective. *Technological Forecasting and Social Change*, 72, 682.
39. FRANTZ, C. M. & MAYER, F. S. 2009. The emergency of climate change: why are we failing to take action? *Analyses of Social Issues and Public Policy*, 9, 205–222.
40. GEELS, F. W. 2005. Processes and patterns in transitions and system innovations: refining the co-evolutionary multi-level perspective. *Technological Forecasting and Social Change*, 72, 682.
41. Ibid.
42. HOFFMAN, A. J. 2010. Climate change as a cultural and behavioral issue: addressing barriers and implementing solutions. *Organizational Dynamics*, 39, 295–305.
43. VAN DER BRUGGE, R., ROTMANS, J. & LOORBACH, D. 2005. The transition in Dutch water management. *Regional Environmental Change*, 5, 164–176.
44. ROTMANS, J., KEMP, R. & VAN ASSELT, M. 2001. More evolution than revolution: transition management in public policy. *Foresight*, 3, 15–31.
45. FRANTZESKAKI, N. & DE HAAN, H. 2009. Transitions: two steps from theory to policy. *Futures*, 41, 593–606.
46. KEMP, R. & LOORBACH, D. *Governance for sustainability through transition management.* Open Meeting of Human Dimensions of Global Environmental Change Research Community, Montreal, Canada, 2003. 12–30.
47. GEELS, F. W. 2005. Processes and patterns in transitions and system innovations: refining the co-evolutionary multi-level perspective. *Technological Forecasting and Social Change*, 72, 682.
48. SEYFANG, G. & SMITH, A. 2007. Grassroots innovations for sustainable development: towards a new research and policy agenda. *Environmental Politics*, 16, 584–603.
49. GEELS, F. W. & SCHOT, J. 2007. Typology of sociotechnical transition pathways. *Research Policy*, 36, 399–417.
50. SEYFANG, G. & SMITH, A. 2007. Grassroots innovations for sustainable development: towards a new research and policy agenda. *Environmental Politics*, 16, 584–603.

51. GEELS, F. W. 2005. Processes and patterns in transitions and system innovations: refining the co-evolutionary multi-level perspective. *Technological Forecasting and Social Change*, 72, 682.

52. SEYFANG, G. & SMITH, A. 2007. Grassroots innovations for sustainable development: towards a new research and policy agenda. *Environmental Politics*, 16, 584–603.

53. BROOK LYNDHURST. 2008. *The diffusion of environmental behaviours: the role of influential individuals in social networks*. Available: http://www.brooklyndhurst. co.uk/the-diffusion-of-environmental-behaviours-the-role-of-influential-individuals-in-social-networks-_110.html (accessed 10 May 2016).

54. VAN DER BRUGGE, R., ROTMANS, J. & LOORBACH, D. 2005. The transition in Dutch water management. *Regional Environmental Change*, 5, 164–176.

55. WENDT, A. 1999. *Social Theory of International Politics*. Cambridge: Cambridge University Press.

56. FRANTZESKAKI, N. & DE HAAN, H. 2009. Transitions: two steps from theory to policy. *Futures*, 41, 593–606.

57. ROTMANS, J., KEMP, R. & VAN ASSELT, M. 2001. More evolution than revolution: transition management in public policy. *Foresight*, 3, 15–31.

58. BÖRZEL, T. & RISSE, T. 2011. From Europeanisation to diffusion: introduction. *West European Politics*, 35, 1–19.

59. WENDT, A. 1999. *Social Theory of International Politics*. Cambridge: Cambridge University Press.

60. PIKE, C., DOPPELT, B. & HERR, M. 2010. Climate communications and behavior change: a guide for practitioners. *The Climate Leadership Initiative* [Online]. Available: http://www.thesocialcapitalproject.org/The-Social-Capital-Project/pubs/climate-communications-and-behavior-change (accessed 10 May 2016).

61. FRANTZESKAKI, N. & DE HAAN, H. 2009. Transitions: two steps from theory to policy. *Futures*, 41, 593–606.

62. PIKE, C., DOPPELT, B. & HERR, M. 2010. Climate communications and behavior change: a guide for practitioners. *The Climate Leadership Initiative* [Online]. Available: https://www.seek.state.mn.us/resource/climate-communications-and-behavior-change-guide-practitioners (accessed 2 June 2016).

63. FRANTZESKAKI, N. & DE HAAN, H. 2009. Transitions: two steps from theory to policy. *Futures*, 41, 593–606.

64. LENZ, T. 2012. Spurred emulation: the EU and regional integration in Mercosur and SADC. *West European Politics*, 35, 155–173.

65. KOTLER, P. & ZALTMAN, G. 1971. Social marketing: an approach to planned social change. *The Journal of Marketing*, 35, 3–12.

66. FRANTZESKAKI, N. & DE HAAN, H. 2009. Transitions: two steps from theory to policy. *Futures*, 41, 593–606.

67. Ibid.

68. SMITH, A., STIRLING, A. & BERKHOUT, F. 2005. The governance of sustainable socio-technical transitions. *Research Policy*, 34, 1491–1510.

69. Ibid.

70. BÖRZEL, T. & RISSE, T. 2011. From Europeanisation to diffusion: introduction. *West European Politics*, 35, 1–19.

71. ROGERS, E. M. 2003. *Diffusion of Innovations*, 5th Edition. New York: Free Press.

72. BUSCH, P.-O. & JÖRGENS, H. 2005. The international sources of policy convergence: explaining the spread of environmental policy innovations. *Journal of European Public Policy*, 12, 860–884.

73. ROGERS, E. M. 2003. *Diffusion of Innovations*, 5th Edition. New York: Free Press.

74. FRANTZ, C. M. & MAYER, F. S. 2009. The emergency of climate change: why are we failing to take action? *Analyses of Social Issues and Public Policy*, 9, 205–222.

75. GEELS, F. W. & SCHOT, J. 2007. Typology of sociotechnical transition pathways. *Research Policy*, 36, 399–417.

76. Ibid.

77. BIO INTELLIGENCE SERVICE. 2012. *Policies to encourage sustainable consumption.* Final report prepared for European Commission (DG ENV). Available: http://ec.europa.eu/environment/eussd/pdf/report_22082012.pdf (accessed 10 May 2016).

78. SPENCE, A. & PIDGEON, N. 2009. Psychology, climate change and sustainable behaviour. *Environment: Science and Policy for Sustainable Development*, 51, 8–18.

79. LENZ, T. 2012. Spurred emulation: the EU and regional integration in Mercosur and SADC. *West European Politics*, 35, 155–173.

80. KOTLER, P. & LEE, N. 2008. *Social Marketing: Influencing Behaviors for Good.* Los Angeles: Sage Publications.

81. BÖRZEL, T. & RISSE, T. 2000. When Europe hits home: Europeanization and domestic change. *European Integration Online Papers (EIoP)*, 4, 1–24.

82. HEINZE, T. 2011. *Mechanism-based thinking on policy diffusion.* KFG Working Paper. Berlin: Freie Universität Berlin.

83. BÖRZEL, T. & RISSE, T. 2000. When Europe hits home: Europeanization and domestic change. *European Integration Online Papers (EIoP)*, 4, 1–24.

84. HEINZE, T. 2011. *Mechanism-based thinking on policy diffusion.* KFG Working Paper. Berlin: Freie Universität Berlin.

85. BÖRZEL, T. & RISSE, T. 2000. When Europe hits home: Europeanization and domestic change. *European Integration Online Papers (EIoP)*, 4, 1–24.

86. PATCHEN, M. 2010. What shapes public reactions to climate change? Overview of research and policy implications. *Analyses of Social Issues and Public Policy*, 10, 47–68.

87. SCHULTZ, P. 2011. Conservation means behavior. *Conservation Biology*, 25, 1080–1083.

88. JACKSON, T. 2005. *Motivating sustainable consumption: a review of evidence on consumer behaviour and behavioural change: a report to the Sustainable Development Research Network.* University of Surrey, Guildford.

89. CHECKEL, J. T. 2005. International institutions and socialization in Europe: introduction and framework. *International Organization*, 59, 801–826.

90. PIKE, C., DOPPELT, B. & HERR, M. 2010. Climate communications and behavior change: a guide for practitioners. *The Climate Leadership Initiative* [Online]. Available: http://www.thesocialcapitalproject.org/The-Social-Capital-Project/pubs/climate-communications-and-behavior-change (accessed 17 May 2016).

91. GIFFORD, R., KORMOS, C. & MCINTYRE, A. 2011. Behavioral dimensions of climate change: drivers, responses, barriers, and interventions. *Wiley Interdisciplinary Reviews: Climate Change*, 2, 801–827.

92. BÖRZEL, T. A. & RISSE, T. 2009. *The Transformative Power of Europe.* The European Union and the Diffusion of Ideas (KFG The Transformative Power of Europe Working Paper Nr. 1, FU Berlin, May 2009).

93. GEELS, F. W. & SCHOT, J. 2007. Typology of sociotechnical transition pathways. *Research Policy*, 36, 399–417.

94. STEG, L. & VLEK, C. 2009. Encouraging pro-environmental behaviour: an integrative review and research agenda. *Journal of Environmental Psychology*, 29, 309–317.

95. GEELS, F. W. & SCHOT, J. 2007. Typology of sociotechnical transition pathways. *Research Policy*, 36, 399–417.

96. BIO INTELLIGENCE SERVICE. 2012. *Policies to encourage sustainable consumption.* Final report prepared for European Commission (DG ENV). Available: http://ec.europa.eu/environment/eussd/pdf/report_22082012.pdf (accessed 10 May 2016).

97. CHECKEL, J. T. 2005. International institutions and socialization in Europe: introduction and framework. *International Organization*, 59, 801–826.

98. BÖRZEL, T. & RISSE, T. 2011. From Europeanisation to diffusion: introduction. *West European Politics*, 35, 1–19.

99. PATCHEN, M. 2010. What shapes public reactions to climate change? Overview of research and policy implications. *Analyses of Social Issues and Public Policy*, 10, 47–68.

100. JACKSON, T. 2005. *Motivating sustainable consumption: a review of evidence on consumer behaviour and behavioural change: a report to the Sustainable Development Research Network*. University of Surrey, Guildford.

101. BÖRZEL, T. & RISSE, T. 2011. From Europeanisation to diffusion: introduction. *West European Politics*, 35, 1–19.

102. BÖRZEL, T. A. & RISSE, T. 2009. *The Transformative Power of Europe*. The European Union and the Diffusion of Ideas, (KFG The Transformative Power of Europe Working Paper Nr. 1, FU Berlin, May 2009).

103. BÖRZEL, T. & RISSE, T. 2011. From Europeanisation to diffusion: introduction. *West European Politics*, 35, 1–19.

104. Ibid.

105. Ibid.

106. ROGERS, E. M. 2003. *Diffusion of Innovations*, 5th Edition. New York: Free Press.

107. GEELS, F. W. 2005. Processes and patterns in transitions and system innovations: refining the co-evolutionary multi-level perspective. *Technological Forecasting and Social Change*, 72, 682.

108. GEELS, F. W. & SCHOT, J. 2007. Typology of sociotechnical transition pathways. *Research Policy*, 36, 399–417.

109. AUSTRALIAN GOVERNMENT PRODUCTIVITY COMMISSION. 2012. *Barriers to effective climate change adaptation*. Available: http://www.pc.gov.au/inquiries/completed/climate-change-adaptation/report (accessed 2 June 2016).

110. UNRUH, G. C. 2000. Understanding carbon lock-in. *Energy Policy*, 28, 817–830.

111. ELZEN, B. & WIECZOREK, A. 2005. Transitions towards sustainability through system innovation. *Technological Forecasting and Social Change*, 72, 651–661.

112. SEYFANG, G. & SMITH, A. 2007. Grassroots innovations for sustainable development: towards a new research and policy agenda. *Environmental Politics*, 16, 584–603.

113. GLOBAL WATER PARTNERSHIP. 2012. *Water Demand Management (WDM) – the Mediterranean experience*. Technical focus paper [Online]. Available: http://www.gwp.org/en/gwp-in-action/News-and-Activities/Global-Water-Partnership-launches-new-publications-at-World-Water-Week-2012/ (accessed 2 June 2016).

114. KEMP, R., SCHOT, J. & HOOGMA, R. 1998. Regime shifts to sustainability through processes of niche formation: the approach of strategic niche management. *Technology Analysis and Strategic Management*, 10, 175–198.

115. GEELS, F. W. 2005. Processes and patterns in transitions and system innovations: refining the co-evolutionary multi-level perspective. *Technological Forecasting and Social Change*, 72, 682.

116. BÖRZEL, T. A. & RISSE, T. 2009. *The Transformative Power of Europe*. The European Union and the Diffusion of Ideas (KFG The Transformative Power of Europe Working Paper Nr. 1, FU Berlin, May 2009).

117. ELZEN, B. & WIECZOREK, A. 2005. Transitions towards sustainability through system innovation. *Technological Forecasting and Social Change*, 72, 651–661.

118. GIFFORD, R., KORMOS, C. & MCINTYRE, A. 2011. Behavioral dimensions of climate change: drivers, responses, barriers, and interventions. *Wiley Interdisciplinary Reviews: Climate Change*, 2, 801–827.

119. MCKENZIE-MOHR, D. 2000. New ways to promote proenvironmental behavior: promoting sustainable behavior: an introduction to community-based social marketing. *Journal of Social Issues*, 56, 543–554.

120. KOTLER, P. & LEE, N. 2008. *Social Marketing: Influencing Behaviors for Good.* Los Angeles: Sage Publications.

121. MCKENZIE-MOHR, D. 2000. New ways to promote proenvironmental behavior: promoting sustainable behavior: an introduction to community-based social marketing. *Journal of Social Issues*, 56, 543–554

122. O'BRIEN, K. L. 2009. Do values subjectively define the limits to climate change adaptation? In: Adger, W. N., Lorenzoni, I. & O'BRIEN, K. L. (eds.) *Adapting to Climate Change: Thresholds, Values, Governance, 164*. Cambridge: Cambridge University Press.

123. MOSER, S. C. & EKSTROM, J. A. 2010. A framework to diagnose barriers to climate change adaptation. *Proceedings of the National Academy of Sciences*, 107, 22026–22031.

124. KOTLER, P. & LEE, N. 2008. *Social Marketing: Influencing Behaviors for Good.* Los Angeles: Sage Publications.

125. DUTCH RESEARCH INSTITUTE FOR TRANSITIONS. 2011. *Urban transition management manual.* Available: https://www.drift.eur.nl/?p=264 (accessed 2 June 2016).

126. ELZEN, B. & WIECZOREK, A. 2005. Transitions towards sustainability through system innovation. *Technological Forecasting and Social change*, 72, 651–661.

127. KEMP, R. & LOORBACH, D. *Governance for sustainability through transition management.* Open Meeting of Human Dimensions of Global Environmental Change Research Community, Montreal, Canada, 2003. 12–30.

128. Ibid.

129. PELLING, M. 2010. *Adaptation to Climate Change: From Resilience to Transformation.* London: Taylor & Francis.

130. ELZEN, B. & WIECZOREK, A. 2005. Transitions towards sustainability through system innovation. *Technological Forecasting and Social Change*, 72, 651–661.

131. ROTMANS, J., KEMP, R. & VAN ASSELT, M. 2001. More evolution than revolution: transition management in public policy. *Foresight*, 3, 15–31.

132. DUTCH RESEARCH INSTITUTE FOR TRANSITIONS. 2011. *Urban transition management manual.* Available: https://www.drift.eur.nl/?p=264 (accessed 2 June 2016).

133 VAN DER BRUGGE, R. & VAN RAAK, R. 2007. Facing the adaptive management challenge: insights from transition management. *Ecology and Society*, 12, 33.

134. VAN DER BRUGGE, R., ROTMANS, J. & LOORBACH, D. 2005. The transition in Dutch water management. *Regional Environmental Change*, 5, 164–176.

135. Ibid.

136. LOORBACH, D. 2010. Transition management for sustainable development: a prescriptive, complexity-based governance framework. *Governance*, 23, 161–183.

137. KEMP, R. & LOORBACH, D. *Governance for sustainability through transition management.* Open Meeting of Human Dimensions of Global Environmental Change Research Community, Montreal, Canada, 2003. 12–30.

138. LOORBACH, D. 2010. Transition management for sustainable development: a prescriptive, complexity-based governance framework. *Governance*, 23, 161–183.

139. DUTCH RESEARCH INSTITUTE FOR TRANSITIONS. 2011. *Urban transition management manual.* Available: http://www.themusicproject.eu/content/transitionmanagement.

140. Ibid.

141. LOORBACH, D. 2010. Transition management for sustainable development: a prescriptive, complexity-based governance framework. *Governance*, 23, 161–183.

142. Ibid.

143. VAN DER BRUGGE, R. & VAN RAAK, R. 2007. Facing the adaptive management challenge: insights from transition management. *Ecology and Society*, 12, 33.

144. Ibid.

145. LOORBACH, D. 2010. Transition management for sustainable development: a prescriptive, complexity-based governance framework. *Governance*, 23, 161–183.

146. KEMP, R. & LOORBACH, D. *Governance for sustainability through transition management.* Open Meeting of Human Dimensions of Global Environmental Change Research Community, Montreal, Canada, 2003. 12–30.

147. DUTCH RESEARCH INSTITUTE FOR TRANSITIONS. 2011. *Urban transition management manual.* Available: http://www.themusicproject.eu/content/transitionmanagement.

148. GIFFORD, R., KORMOS, C. & MCINTYRE, A. 2011. Behavioral dimensions of climate change: drivers, responses, barriers, and interventions. *Wiley Interdisciplinary Reviews: Climate Change*, 2, 801–827.

149. Ibid.

150. KEMP, R. & LOORBACH, D. *Governance for sustainability through transition management.* Open Meeting of Human Dimensions of Global Environmental Change Research Community, Montreal, Canada, 2003. 12–30.

151. ELZEN, B. & WIECZOREK, A. 2005. Transitions towards sustainability through system innovation. *Technological Forecasting and Social Change*, 72, 651–661.

6 Transitions towards managing natural resources and water

Introduction

In natural resource management, there are two types of transitions society can undergo in managing scarce resources sustainably: non-transformative and transformative transitions. In non-transformative transitions, institutions inform individuals and communities of the need to conserve scarce resources, while transformative transitions involve individuals and society adjusting their attitudes and behaviour towards the environment and its natural resources.[1] An important distinction is made between the two with transformative approaches radically re-imagining people's relationship with the environment, while non-transformative methods are conventional, limited in scope and only aspire to achieve small, incremental change.[2] As such, transitions towards sustainability are transformative in that institutions attempt to change the attitudes and behaviour of individuals and society in order to achieve a specific sustainability target.[3,4] However, it is a public choice as to how fast they make a transformative change towards managing natural resources sustainably. In particular, society has two choices: first, an orderly transition, on the public's terms, or, second, a disorderly transition in which environmental degradation dictates the speed and timing of the transition. If the transition is dictated by the environment, it is likely the transition will be unacceptably harsh for humans to adjust to, in terms of structural change and cost.[5]

This chapter will first introduce transitions in natural resource management that are driven by climatic and non-climatic drivers before discussing the transition framework in the context of urban water management. The chapter will then discuss how transitions in urban water management can be operationalised and assessed, before finally discussing the numerous barriers that can exist in transitions towards urban water security.

Urban Water Security, First Edition. Robert C. Brears.
© 2017 John Wiley & Sons, Ltd. Published 2017 by John Wiley & Sons, Ltd.

6.1 Transitions in natural resource management

In natural resource management, drivers of transitions can be grouped into climatic and non-climatic drivers. Regarding climate change drivers, there are two approaches for individuals and society to adapt to the pressures of climate change: mitigation and adaptation. Traditionally, it is common for local authorities to mitigate the impacts of climate change by taking actions that prevent the impact of an event, for example, the construction of coastal seawalls to protect communities from sea-level rise and increased storm surges as a result of climate change. However, these 'hard' infrastructural solutions are costly both economically and environmentally to implement.[6,7]

6.1.1   Adaptation towards climate change

Adaptation towards climate change is an ongoing process in which actors seek to reduce the vulnerability or enhance the resilience of a society towards observed or expected climate change.[8,9] Specifically, adaptations within social groups involve attempts at reducing the impacts of climate change on individuals' livelihoods and society's well-being.[10] Adaptations occur over multiple dimensions including spatial (local, regional, national) and temporal (responses to current vulnerability, observed medium- to long-term trends in climate and anticipatory in response to model-based scenarios of long-term climate change), occur across many sectors (water), involve numerous actions (physical, technological, investments, regulations and markets), occur over many climatic zones (temperate, tropical) and involve many actors (local authorities, government, public and private sectors, communities and individuals).[11,12] The aim of adaptation is to increase the adaptive capacity of a system to successfully respond to climate change through behavioural, resource and technological adjustments and reduce the risks associated with the impacts of climate change.[13,14] There are numerous considerations regarding the implementation of adaptation[15]:

1 *Adaptation is location and context specific*: There is no single approach to reducing risks appropriate across all settings. Effective adaptation strategies consider local vulnerabilities and exposure to risks and their linkages with socioeconomic factors, sustainable development and climate change.
2 *Adaptation can be enhanced by multilevel governance*: Adaptation planning and implementation can be enhanced through complementary actions across multilevels of governance. National governments can coordinate adaptation efforts of both local and sub-national governments by providing information, policy and legal frameworks and financial support. Local governments, with the private sector, are becoming recognised as being critical to the adaptation process particularly due to their ability to scale up adaptation of communities, households and civil society and provide information and financial resources.
3 *Reducing vulnerability and exposure to present climate variability*: Resilience of individuals, communities and populations to future climate change variability can be increased by improving current levels of human health, livelihoods, social and economic well-being and environmental quality.

4 *Diverse values, objectives and risk perceptions need to be incorporated*: Local and traditional knowledge systems and practices are an important resource for adaptation to climate change; however, they have not been frequently incorporated into existing adaptation efforts. Integrating local knowledge with existing practices can increase the effectiveness of adaptation.

5 *Adaptation is more effective when information is context driven*: Organisations that bridge science and decision-making play an important role in communicating, transferring and developing climate-related knowledge.

6 *Existing or new economic instruments can foster adaptation*: Economic instruments, both current and new, can foster adaptation by providing incentives for anticipating and reducing impacts. Instruments can include resource pricing, subsidies, regulations and financial loans.

7 *Barriers can impede adaptation*: Barriers can be external such as financial, economic, capacity and so on and internal including differing perceptions of risk and competing values.

8 *Poor planning can result in inadequate adaptation*: Lack of planning in adaptation can increase the vulnerability or exposure of target groups in the future or increase the vulnerability of people, places or sectors.

6.1.2   Types of adaptations: Green and soft

There are two main types of actions that can be taken in adaptations to climate change in natural resource management: green actions and soft actions. Green actions ensure ecosystem health is maintained by ensuring natural resources are used as efficiently as possible reducing society's vulnerability to risks from scarcity.[16] Green actions are usually less resource intensive than mitigation (hard actions) in terms of financial and technical capacity, as green actions do not require the development and maintenance of large-scale infrastructure.[17] Nonetheless, green actions frequently overlook the social dimensions of climate and environmental change; instead they focus on economic and technological solutions to the problems.[18] In soft actions, the focus instead is on using management and legal and policy approaches to alter human behaviour as a way of reducing vulnerability to climate change risk.[19] In addition, soft actions attempt to reduce society's impact on environmental degradation by decoupling resource consumption from economic and population growth.[20]

6.1.3   Managing resource scarcity

In natural resource management, institutions seek to reduce the vulnerability of society from environmental degradation and resource scarcity, as a result of urbanisation and population growth, by transitioning from a first-order scarcity sociotechnical system to eventually a third-order scarcity sociotechnical system.[21]

In first-order scarcity, institutions rely on mitigation as a way of meeting actual or perceived supply inadequacies. In particular, natural resource managers address first-order scarcity by constructing large-scale infrastructural projects to increase supply. Because of the large economic and environmental costs associated with supply-side projects, natural resource managers have turned to second-order scarcity policies, which focus on improving economic and technological efficiency in managing demand and supply of natural resources.[22]

In second-order scarcity, adaptations involve economic and technological inputs to manage natural resources more efficiently. However, while economic instruments and technological developments may provide solutions to environmental degradation, individual beliefs, norms and values drive environmental change.[23] As such, in order to properly address environmental degradation, there needs to be a transition in societal values, in particular changes in behavioural patterns, thinking and value structure towards the environment, so that society recognises that climate change and environmental degradation are not only a scientific fact but also a social fact.[24,25]

In third-order scarcity, the focus is on behavioural change as a way of decreasing demand for scarce natural resources, which in turn lowers environmental degradation. In particular, third-order scarcity, while incorporating economic and technological solutions, focuses on soft actions that seek to reduce demand for resources by altering the individual's and society's norms and values towards the environment and its resources. The reasoning is that both environmental problems and solutions are culturally rooted.[26,27,28] In third-order scarcity, the altering of norms and values can be instigated through the promoting of pro-environmental behaviour, which is behaviour that has as little environmental impact as possible or is beneficial to the environment.[29] Combined, natural resources can be managed in a way that adapts to both climate change and resource scarcity (Table 6.1).

Table 6.1 Managing the impacts of climate change and resource scarcity

Adaptation type	Level of scarcity	Role of demand	Dominant discipline and response
Mitigation	First	Forecasted based on historical records	Hard infrastructural projects
Green adaptation	Second	Projections based on economic variables	Economic demand-side management. Resources are economic goods and the object is to increase economic and technical efficiency
Soft adaptation	Third	Scenarios based on economic, demographic and social variables	Managing demand for resources through altering of behaviour

WOLFE, S. & BROOKS, D. B. 2003. Water scarcity: an alternative view and its implications for policy and capacity building. *Natural Resources Forum*, 27, 99–107

6.2   What is a transition in urban water management?

A transition in urban water management is a well-planned, coordinated transformative shift from one water system to another, over a long period of time (usually one or two generations), where a water system is composed of physical and technological infrastructure, cultural/political meanings and societal users (Table 6.2).[30,31] In a water system, society is both a component of the water system and a significant agent of change in the system, both physically (change in processes of the hydrological cycle) and biologically (change in the sum of all aquatic and riparian organisms and their associated ecosystems).[32] More specifically, a transition from one water system to another involves a structural change in the way society manages its scarce water resources and occurs through a combination of behavioural, cultural, ecological, economic, institutional and technological developments that positively reinforce each other to create a new water system.

6.2.1   Drivers of transitions in urban water management

A transition towards a new water system is triggered by changes in the external environment of the system, leading it to being inefficient, ineffective or inadequate in fulfilling its societal function: ensuring urban water security for all users. The main drivers in transitions towards new sociotechnical systems in urban water are rapid population growth and demographic changes, rapid urbanisation, rapid economic growth and rising incomes, increased demand for energy and food and climate change. The impacts of each driver on water resources are listed in Table 6.3.

Table 6.2 Components of a water system

Sociotechnical system for water supply

Regulations and policies	Managing the quality and quantity of water resources
Infrastructure	Drinking, storm and wastewater network
Treatment	Drinking and wastewater
Markets and users	Domestic and non-domestic users' habits, expectations and practices
Drinking water	Quality of supply
Culture	Cultural and symbolic meanings (social and cultural values of water, the use of water and water technologies)

GEELS, F. W. 2005. Processes and patterns in transitions and system innovations: refining the co-evolutionary multi-level perspective. *Technological Forecasting and Social Change*, 72, 682

SOFOULIS, Z. 2005. Big water, everyday water: a sociotechnical perspective. *Continuum: Journal of Media & Cultural Studies*, 19, 445–463

Table 6.3 External drivers of transitions in IUWM

Driver	Impact
Population growth and demographic changes	Rapid population growth increases demand for water, for both domestic and non-domestic use, leading frequently to over-exploitation of water resources. This results in excessive withdrawals and water scarcity. Meanwhile, demographic changes lead to change in consumption patterns impacting water quantity and quality
Rapid urbanisation	Urbanisation (urban sprawl or encroachment into river basin catchment areas) lowers the availability of good quality water of sufficient quantity through point source pollution (industrial, domestic wastewater) and non-point source pollution (pathogens, organic and inorganic). Over-exploitation of groundwater and surface water degrades ecosystems and their services (e.g. reduced ability to purify water)
Economic growth and rising income	Competing water use can lead to over-exploitation resulting in inter-sectoral, inter-regional and even international competition over scarce water resources. Rising income levels will increase household demand for water, while diets will shift to water-intensive products
Energy demand	Water and energy are linked in two ways: first, water is used in the production of almost every type of energy (coal, geothermal, hydro, oil and gas, nuclear and biofuel) and, second, energy is a dominant cost factor in providing water and wastewater services
Food demand	With rising populations and changes in diet, demand for food is increasing resulting in the need for additional water resources. At the same time, nutrient runoff from agricultural production impacts water quality, degrading ecosystems
Climate change	Storm events (flooding) wash pollutants from urban areas into surface water bodies as well as contaminate groundwater supplies. As urban populations encroach into river basins, they are at increased risk of contaminated water supplies during flooding events
	Built environments, including buildings and roads, absorb sunlight and re-radiate heat. This combined with less vegetative cover, which provides shade and cools moisture in the air, means air temperatures of urban areas are 3.5 to 4 degrees Celsius higher than surrounding rural areas. The result is an increase in demand for water for cooling and drinking
	During heatwaves and droughts demand for water increases (drinking water and water for cooling). In addition, with increased temperatures, oxygen levels in water will decrease, while algal levels will increase, degrading the quality of water resources leading to increased treatment costs and energy use in the treatment process
	Globally, cities are mainly concentrated in coastal zones resulting in a large portion of the world's urban population exposed to the risk of sea-level rise and intensifying storm surges, which contaminate groundwater supplies and damage water infrastructure

6.2.2   Transitioning from supply-side to demand-side management

In traditional water resource management (first-order scarcity), urban water managers forecast population growth and economic development to determine future levels of demand. If there is a projected supply deficit (demand outstripping supply), traditional water management relies on large-scale water supply projects consisting of dams, reservoirs and pipelines to transport water over large distances to bridge that gap.[33,34,35,36,37,38,39] Over time, however, these supply-side solutions have become unfavourable due to their environmental, economic and political costs. Environmentally, supply-side solutions such as dams and reservoirs impact the quantity and quality of water available for ecosystems, impacting adversely the numerous ecosystem services both humans and nature rely on for their survival.[40,41,42,43,44,45] There are large economic costs too involved with supply-side solutions, in particular, the reliance on more distant water, often of inferior quality, to meet rising demand has not only increased the costs of transportation (energy costs) but also treatment costs (chemical costs).[46,47] Politically, scarcity of water is likely to lead to inter-user, inter-sectoral, inter-regional and international competition, or even conflict, over scarce water resources, as aquifers, rivers, lakes and even entire river basins frequently cross internal and external political boundaries. Without ongoing dialogue and cooperation between municipalities, regions and states, unilateral actions, for example, water abstractions, can lead to significant impacts on water resources in neighbouring local, regional or national jurisdictions.[48,49,50] In addition to these costs, traditional water resource management fails to account for uncertainty in supply from climate change extremes (floods and droughts) and changing weather patterns (spatial and temporal changes in precipitation).[51,52] As such, with increased demand for water from urbanisation and variability of supply from climate change, traditional water management practices have become outdated.[53]

In second-order scarcity, water managers explore demand-side options in the management of scarce water resources. Specifically, rather than projecting current demand trends forwards and then trying to find the water to meet those needs, water managers deconstruct demand to determine actual needs and the most efficient ways of meeting those needs.[54] To ensure water is used in the most efficient way, second-order policies focus on increasing economic and technical efficiency in water use. In particular, attention is paid to the economic value of water which encourages the introduction of pricing of water to end users and the subsequent need to meter water consumption.[55] However while second-order scarcity policies may be sufficient for a few years, at some point they have to give way to third order policies as a result of water scarcity from climate change and urbanisation.[56,57]

Third-order scarcity policies are directed at shifting the emphasis away from economic and technical efficiency towards addressing the actual driver of water demand: human behaviour. Specifically, third-order scarcity policies combine second-order scarcity of economic and technical efficiency with demand management which is a process in which ideas, norms and innovations of water conservation are communicated across water users in a community, the purpose being to change people's attitudes, culture and practices towards water and reduce consumption patterns in order to achieve a targeted level of water consumption.[58,59] To decrease demand for water through social change in third-order scarcity, water managers, first, examine how identities (behaviours, norms and values) are formed, maintained and modified and, second, define a future ideal level of water consumption and work backwards to find a feasible and desired pathway of changing people's attitudes and behaviour towards water to achieve that vision.[60,61] The eventual goal of third-order scarcity in urban water management is to decouple water consumption from economic and population growth and achieve water security for all users and uses.[62]

6.2.3   *Types of transitions in third-order scarcity*

There are four types of transformative transitions that can occur in transitions towards third-order scarcity: endogenous, reorientation of trajectories, emergent transformations and purposeful transformations. In endogenous transformations water managers base their strategic visions on perceived pressures to water supply such as water scarcity and economic competitiveness from other cities. Reorientation of trajectories result from shocks such as droughts impacting the availability of good quality water of sufficient quantity, while emergent transformations arise from uncoordinated pressures outside of the regime which can include limited groundwater supplies from contamination, increased costs of energy and political costs of importing transboundary water. Finally, purposeful transformations are intended and coordinated with water managers fulfilling the desire of society to manage water and economic resources sustainably.

6.3   Operationalising transitions in third-order scarcity

In transitions towards third-order scarcity, there are two components to modifying the attitudes and behaviour of water users (domestic and non-domestic) at the meso level: first, there is the strategic or macro-level sustainability vision or goal – the water-saving target – and, second, the operationalisation of this strategy at the micro level.[63] In transitions towards third-order scarcity, the application of supportive forces at the macro level can be in the form of targeted levels of water consumption (e.g. per capita litres per day) with the baseline for comparison being current levels of (unsustainable) water consumption. At the micro level, using the definition of diffusion, the application of formative forces can be in the form of demand management. Demand management, in IUWM, is a process in which ideas, norms and innovations of water conservation are communicated across water users in a community. Its purpose is to change water user's culture, attitudes and practices towards water and reduce consumption patterns in order to achieve a targeted level of water consumption.

6.3.1   *Setting the macro-level strategic goal*

At the macro level, water managers set water conservation goals where water conservation is defined as the beneficial reduction in water use, waste and loss.[64] As part of the macro-level vision, it is important for water managers to convey to water users how much water could be saved from water conservation programmes as the success of water conservation efforts depends on public awareness and the understanding of the need to conserve water.[65] In particular, water managers need to convey the multiple benefits that arise from conserving water which include reductions in energy consumption, which in turn saves money on water and energy bills; reductions in the amount of sewage that requires treatment, which extends the 'life' and capacity of sewage treatment plants, reducing the need to construct new sewage treatment plants or expand existing plants to meet increased demand for water resources in the future; and reductions in water use that can delay or prevent costly expansions of the water supply system, which has large capital costs from water treatment plants, reservoirs, distribution pipelines and so on.[66] Nonetheless, water managers need to recognise that average household water consumption rates represented by average litres per day do not reflect diversity of water using practices and activities by individuals and households, such as cleanliness and comfort (laundry, washing), gardening and forms of pleasure, for example, drinking and eating within homes, nor does it represent how socio-demographic variables (e.g. age, gender, ethnicity, house ownership and type, etc.) impact water usage.[67]

In any case, individuals should gain the following understandings in a water conservation programme: the environmental benefits of lowering water demand, in particular how excess demand reduces groundwater and surface water supply impacting the health of ecosystems; water conservation helps water quality in that more water is retained by the natural environment, diluting non-point source and point source pollution as well as providing cooler temperatures for aquatic wildlife; investments in efficiency and conservation will provide water users with long-term savings, compared to the extra costs of developing new water supply sources and wastewater treatment plants; conservation helps reduce the costs of providing water (operation and maintenance of water supply systems, treatment costs of providing water and treating wastewater); and new water-saving devices and technologies can increase the efficiency of water use.[68,69]

6.3.2   Micro-level demand management tools

At the micro level, demand management in third-order scarcity comprises a set of policies that promote the better use of existing urban water supplies before plans are made to increase supply. Using the rationalist/constructivist approach in diffusion, water managers can use two types of demand management strategies to influence the norms and values of society towards water resources: antecedent and consequential strategies. Antecedent strategies attempt to influence the determinants of target behaviour through activities such as increasing individuals' knowledge or awareness of problems through information campaigns, behavioural commitments and prompting. The assumption is that these strategies can influence the determinants of behaviour before its performance. Consequential strategies (feedback, rewards and punishments) are all assumed to influence the determinants of target behaviour *after* the performance of the behaviour. The latter strategy assumes that feedback, both positive and negative, of the consequences of that behaviour influences the likelihood of that behaviour being performed in the future.[70]

Water managers can apply antecedent and consequential strategies in the form of direct and indirect demand management tools: direct demand management tools attempt to modify water users' attitudes and behaviour towards water resources through manipulation of utility calculations, legal or physical coercion, socialisation and persuasion (Table 6.4), while indirect demand management tools of competition, lesson drawing and emulation and mimicry are used as prompts by water managers to promote water conservation (Table 6.5). In demand management, water managers typically start with pilot programmes designed to identify the best approaches of promoting conservation efforts in a specific community before scaling up to more expensive, broader efforts.[71]

Table 6.4 Direct demand management tools

Direct demand management tools	Type of tool	Description
Manipulation of utility calculations	Regulatory and technological	Water pricing can be used as an incentive to increase water efficiency and promote water conservation. In particular, water pricing internalises the environmental and social costs of water use (in addition to raising revenue for the operation and maintenance of water supply infrastructure)
Legal or physical coercion	Regulatory and technological	Water bans or water restriction, rules and regulations in homes and commercial buildings for water efficiency
Socialisation	Regulatory and technological	Water managers can promote water conservation through the use of authoritative schemes such as labelling, accreditation and certification of water efficiency in appliances, building designs and so on
Persuasion	Communication and information	Education programmes in schools can be also used to persuade young people to conserve water resources. Water managers can use public education to persuade individuals to conserve water. This can be conducted through various multimedia formats (TV, radio, newspapers, Internet, etc.)

Table 6.5 Indirect demand management tools

Indirect demand management tools	Type of tool	Description
Competition	Communication and information	Water managers can promote competition between water users by enabling the comparison of one's own water consumption or savings with the average water consumption or savings of others
Lesson drawing	Communication and information	Water managers can provide individuals and communities with information on water conservation practices that have worked elsewhere and are easily transferable to the local context
Emulation and mimicry	Communication and information	Water managers can promote individuals or communities that have made considerable water savings as a standard for others to emulate. Similarly, water managers can provide tips on how to mimic others' water savings. Water utilities can enact corporate social responsibility plans to reduce water and energy use as well as carbon emissions

6.3.3   Transition management cycle in third-order scarcity

Using the transition management cycle, urban water managers can plan, implement, monitor and evaluate transitions towards third-order scarcity. This enables urban water managers to learn from experiments and constantly assess and periodically adjust policies for achieving the vision.[72,73,74] In particular, urban water managers can, first, define the problem and create a vision of an alternative water future. Second, they can develop an agenda to achieving that vision by selecting a target behaviour that has the largest impact on achieving the vision and identify the barriers to achieving it. Third, they can establish and conduct experiments that provide new solutions and instruments to achieving the vision. Last, they can monitor and evaluate the experiments regularly and learn before adjusting, where necessary, the vision and agenda. If experiments are successful, as defined by its contribution towards achieving the transition vision, they can be repeated in different contexts (broadened) and scaled up (deepened) from the micro to macro level.[75,76,77,78]

6.3.4   Analysing transition management cycles: SWOT analysis

The analysis of transition management cycles can be done using SWOT analysis technique, an acronym that stands for strengths (S), weaknesses (W), opportunities (O) and threats (T). SWOT analysis is based on the concept of 'strategic fit': the idea that an organisation is successful if its internal characteristics (strengths and weaknesses) fit the external environment (opportunities and threats) with the fundamental role of strategic planning being to ensure this fit in the long run. Once the internal and external factors have been identified, SWOT pairs the strengths and weaknesses with the opportunities and threats, giving four types of strategic possibilities: using strengths to exploit opportunities, using strengths to avoid or minimalise threats, identify and address weaknesses that may prevent achieving objectives and identify weaknesses that make the organisation vulnerable to threats.[79] Nonetheless, these possibilities are not necessarily alternatives; instead the key is to develop a strategy that combines these possibilities in a way that ensures the 'strategic fit'; therefore the organisation is able to maximise benefits that come from the changing environment. The subject of a SWOT analysis can vary as the technique is flexible: it can be applied to a company, a policy or even a development programme. It can also be used to judge the strategic fit in relation to a particular objective: whether the objective is attainable or not.[80]

6.4   Barriers to transitions towards urban water security

In transitions towards urban water security, even the most planned transitions cannot be barrier-free, where barriers are defined as anything that prevents the community from using its resources in the most advantageous way in responding to climate and environmental change.[81,82,83,84] Similar to the process of diffusion, there are multiple barriers, both external and internal, to the introduction and diffusion of pro-environmental innovations.[85,86,87] In transitions towards urban water security, these barriers create gaps between attitudes and behaviour towards the environment.[88]

6.4.1 External barriers

Economic and financial

Diffusions of innovations face numerous economic and financial barriers.[89,90,91,92] Regarding economic barriers, many new practices (social and behavioural) and technologies have failed to become mainstream due to the lack of economies of scale. Without scale, new products and techniques cannot compete on price, and therefore innovations are not taken up by individuals.[93,94] Meanwhile, the lack of financial capital available for individuals and businesses stifles the development of innovative practices such as alternative water supplies that can reduce demand for potable water,[95,96] resulting in current technology being locked in: made worse with the global financial crisis straining local authorities' budgets.[97,98,99,100,101,102,103] For example, in the United Kingdom there is a lack of grants or subsidies for feasibility assessments that provide non-product support to stakeholders interested in implementing rainwater harvesting systems.[104] Institutions also create financial barriers to the diffusion of innovations by tying the success of innovation with further funding; however, this does not allow niches to adapt and overcome problems through experimentation, reducing the volume of innovations that will become mainstream in the future.[105] Diffusions of innovations also face market barriers, which are conditions that prevent the markets from allocating resources to users and uses efficiently. Market failures can occur due to inadequate information on climate change reducing the ability of households, businesses and governments to make well-informed decisions.[106,107] In addition, water utilities may face financial limitations in developing new water conservation programmes, training staff or hiring additional staff to implement new programmes, as well as educating individuals and communities on water conservation.[108]

Environmental

Environmental barriers are thresholds, or limits, beyond which existing activities, land uses, ecosystems, species or system states cannot be maintained naturally or even artificially.[109,110,111] The actual limit is dependent on the rate and magnitude of change and what the tolerance of that particular system is, beyond which the system may not be able to adapt.[112] However, water managers often lack adequate climate and hydrological information needed to plan, develop and manage water sustainably to ensure environmental limits are not reached.[113]

Infrastructure and technology

A lack of appropriate infrastructure can impede the development of innovations with current infrastructure (technologies and systems) being unable to support alternative practices: a simple example being people unable to recycle household materials due to there being no recycling centres to process the material.[114,115,116,117,118,119] Often this is due to relying on conservative, highly visible infrastructure solutions rather than attempting to do new things.[120] However, without economies of scale for new innovations, there is often no profitability in constructing new infrastructure to support alternative practices as the cost of building new infrastructure outweighs the fixed cost of operating the existing infrastructure.[121] In addition, new technology may not fit well into the existing system. In particular, new technologies may require complementary technologies that may not be available or are expensive or difficult to use. Specifically, new innovations may either be expensive because of a lack of economies of scale in reducing production costs, or they are difficult to use (because the innovation has not been tested by customers on a large scale and, therefore, needs redesigning before it can be mainstreamed or found to be culturally desirable.)[122,123,124]

Institutional

The extent that adaptation strategies are implemented is dependent on the capacity of institutions (institutional capacity) and individuals (human capacity) to implement the strategies and willingness of individuals to adopt the strategies.[125,126,127,128] Lack of institutional capacity can result in sub-optimal outcomes due to institutional managers being unable to link micro-level projects with macro-level visions. It can also lead to policymakers becoming too risk adverse in initiating best practices that worked elsewhere unless there is strong empirical evidence to suggest otherwise.[129,130]

Political

Institutions often create barriers to diffusion due to institutions lacking leadership or political will to initiate and sustain a transition.[131,132,133,134,135,136,137,138,139,140,141] For instance, water utility managers may lack significant support from superiors to initiate water conservation measures. If political leadership is lacking, then water utilities could fail in coordinating education and outreach between water utilities and industry (developers, contractors, investors, suppliers, etc.).[142] Lack of political leadership or political will is often due to the lack of defined responsibility for decision-making or leadership lacking quality (skill set), integrity, transparency and accountability, coordination/interaction between government bodies or capacity (financial and technical) in managing natural resources across various sectors of the economy and society.[143,144,145,146]

Regulatory

Institutions often impede innovations through various regulations, from which they become an independent force assuming a life of their own. The result is sub-optimal outcomes adhering to standards/regulations rather than the implementation of more efficient, or optimal, choices that violate the standards.[147,148,149,150,151,152,153]

Regulatory barriers can be in the form of existing standards stifling innovations, regulatory costs and failure of institutions to use regulations to promote innovations. Frequently, existing regulations form a barrier to the development of new innovations since the regulations are usually created for an existing regime. Therefore, it may be problematic or even illegal to work on a novelty in the public domain.[154,155,156,157] Existing regulations can enforce rigidity through fixed investments in technology/infrastructure and competency.[158,159] Regulatory costs can stifle innovations by making it costly to implement new innovations. In particular, before innovations can be implemented, there may be compliance costs borne by households and businesses in meeting regulations or administrative costs borne by the government administering the regulation or by households and businesses in paperwork, reporting time, fees and charges imposed by regulators.[160] Finally, institutions often fail to clearly signal the need for specific new technologies, and so there is a lack of subsidies, support and policies to develop new technologies.[161]

6.4.2   Internal barriers

In diffusion there are two types of internal barriers: psychological and social.[162] Psychological barriers exist due to the desire of humans to have ontological security, in particular order and predictability,[163,164] while social norms and values (including culture, perceptions, customs, knowledge, traditions and levels of cognition) act as barriers to diffusion by ensuring prevailing activities are deemed satisfactory, ensuring current practices continue into the future.[165,166,167,168,169,170]

6.4.3   Psychological barriers

Knowledge/information/awareness

Numerous studies have found that large proportions of people in numerous countries have little awareness, knowledge or understanding of the importance of climate change, how global change can have local impacts and how their personal decisions impact the environment. For instance, it is common for individuals to lack knowledge of the city's water supply system and where water is sourced.[171] This lack of information and knowledge directly impacts the likelihood of individuals taking pro-environmental actions as evidence shows that as knowledge of the environment increases, people's willingness to take positive environmental action increases.[172,173,174,175,176,177,178,179,180,181] To increase knowledge and awareness on climate change and sustainability, most programmes that aim to foster pro-environmental behaviours have been information intensive with common mediums being media advertising and distribution of printed materials used to foster behavioural change.[182] Information-based campaigns are usually based on two perspectives of behavioural change: first, programme planners assume that by enhancing knowledge on environmental issues, it will encourage the development of attitudes that are supportive of the activity. Second, programme planners assume that individuals act rationally and so focus their information campaigns on the economic advantages of engaging in specific pro-environmental activities.[183] However, simply providing people with environmental information does not always translate into pro-environmental actions because of the following reasons: First, people frequently lack basic knowledge on how the world's physical environment works, for example, many people are unfamiliar with how to implement water conservation techniques.[184] Second, individuals often find it difficult to process multiple sources of information and so information overload can stifle pro-environmental behaviour. Third, terminologies such as 'sustainability' and 'sustainable development' can mean different things to different people, creating a significant barrier to learning on how to become sustainable.[185,186,187,188,189,190,191,192]

Lack of connection with nature

People's values are frequently shaped by their self-image and so they act in ways that enhance their image. Studies have revealed that the more a person perceives their self-image to be connected with nature, the more pro-environmental actions the person will take to preserve nature.[193,194,195] However, as populations have become more urbanised, the portion of time people spend in nature has decreased significantly, resulting in people having a lack of familiarity with nature.[196,197] As such, there is often a lack of understanding between today's water use and consequences in terms of costs and resource quality.[198]

Uncertainty/scepticism towards climate change

Humans frequently fail to act on climate change and environmental issues for three main reasons: first, humans are often unable to perceive slow incremental changes and therefore are less likely to modify their behaviour for a threat they cannot see; second, climate change and environmental problems are complex; however, humans often tend to simplify issues, which leads to the loss of deeper understandings of the consequences of human behaviour and underestimating of the extent of the problem; and third, people are uncertain about the existence of climate change and this uncertainty provides justification for inaction or postponed action.[199,200,201,202,203]

Distrust in information sources

Studies have revealed that people fail to take action on climate change and environmental problems because they believe the media exaggerates, or sensationalises, climate change stories. As such, research has shown the more trusted the source, the more likely individuals are to take positive environmental actions.[204,205]

Fear framing and denial/lack of action/fatalism

There is an assumption in environmental public policy that when people are provided with information, they will naturally change their behaviour in an environmentally beneficial way; however, frequently people fail to do so.[206] This is because environmental messages are frequently framed around the use of fear.

While fear of climate change and environmental degradation can influence pro-environmental behaviour, there is, however, a limit to the amount of fear that can be invoked in messages that aim to modify behaviour. When too much fear is applied, rather than initiate pro-environmental actions, it can lead to inaction, feeling of hopelessness and even the denial of the threat's existence, resulting in pro-environmental inaction.[207,208,209,210]

Technology will solve all problems

Research has found the more over-confident people are in technology solving climate change and environmental degradation, the less pro-environmental actions they will take in reducing their own personal impact on the environment. This ignores the fact that humans and their economic, social and cultural values are the drivers of environmental degradation.[211,212,213,214]

Climate change is a distant threat

It is common for people to view climate change and environmental degradation as a distant threat spatially and temporally[215]: spatially, climate change only impacts remote areas of the world, for example, in the Arctic, while temporally climate change will only happen 'in the future sometime'.[216,217,218,219,220,221] Combined, individuals fail to connect their personal consumption choices and behaviour with climate change and environmental degradation, resulting in less motivation to improve their local environments.[222,223]

Other issues are more important

Because most people fail to understand how climate change and environmental degradation can directly threaten property and life, they place higher importance on other issues such as poor economic growth, personal issues, unemployment and even transportation.[224,225,226]

Reluctance to change lifestyles

Research has found that people will take pro-environmental actions when it does not seriously decrease their lifestyle or quality of life.[227] The reason is that people often view sustainable lifestyles and pro-environmental behaviour as less fun, progressive, advanced or developed and subsequently of lower quality in comparison with others who are not acting sustainably.[228,229,230,231,232] Therefore, the greater the personal sacrifice, the less pro-environmental actions individuals will take.[233] In addition, when it comes to sustainable technologies, it is common for people to believe they are costly and expensive, and therefore their uptake will decrease standards of living and quality of life.[234,235]

Drop in the ocean feeling

When people are confronted by large-scale issues, such as climate change and environmental degradation, their motivations to act pro-environmentally (reduce their personal contribution to the problem) depend on their perceptions of whether action taken by themselves can make a difference or not.[236] Specifically, the more people feel their behaviour will not make a difference, the less likely they are to act pro-environmentally.[237,238,239,240]

6.4.4   Social barriers

Lack of action by business and government

Research reveals that many people believe it is businesses' and the government's responsibility to address climate change. When people perceive businesses and governments as not doing enough, it is an indication that the problem is not as large or urgent as it is made out to be or that solutions are unavailable. Therefore it will be 'business as usual'. In response, individuals will not feel morally obliged to change their behaviour to reduce their environmental impacts.[241,242,243,244,245]

Worry about free rider effect

People's feelings of personal responsibility towards protecting the environment are frequently influenced by the actions of others. In particular, people are less willing to translate their concern for the environment into pro-environmental behaviour if they perceive other individuals to be unconcerned or unwilling to take pro-environmental actions.[246,247,248]

Different demographic groups

Institutions often fail to mainstream pro-environmental behaviours because of a lack of understanding of the various demographic groups that comprise society. In particular, programme managers often fail to determine what pro-environmental behaviours specific demographic groups already engage in and how additional pro-environmental behaviours can be promoted based on each group's unique experiences, knowledge, interpretations and responses to climate change, which are dependent on each group's world views, values, identities and beliefs.[249,250,251,252] For instance, water-using practices of individuals and households change over time due to changes in work and leisure practices (e.g. working more from home or more active in sports), personal hygiene views on what is 'presentable' and family structure with children growing up and socialising outside more often.[253]

Notes

1. VALLANCE, S., PERKINS, H. C. & DIXON, J. E. 2011. What is social sustainability? A clarification of concepts. *Geoforum*, 42, 342–348.
2. Ibid.
3. DUTCH RESEARCH INSTITUTE FOR TRANSITIONS. 2011. *Urban transition management manual* [Online]. Available: https://www.drift.eur.nl/?p=264 (accessed 3 June 2016).

4. SEYFANG, G. & SMITH, A. 2007. Grassroots innovations for sustainable development: towards a new research and policy agenda. *Environmental Politics*, 16, 584–603.

5. GOODLAND, R. 1995. The concept of environmental sustainability. *Annual Review of Ecology and Systematics*, 26, 1–24.

6. AUSTRALIAN GOVERNMENT PRODUCTIVITY COMMISSION. 2012. *Barriers to effective climate change adaptation* [Online]. Available: http://www.pc.gov.au/__data/assets/pdf_file/0008/119663/climate-change-adaptation.pdf (accessed 3 June 2016).

7. EUROPEAN ENVIRONMENTAL AGENCY. 2013. *Adaptation in Europe – addressing risks and opportunities from climate change in the context of socio-economic developments* [Online]. Available: http://www.eea.europa.eu/publications/adaptation-in-europe (accessed 11 May 2016).

8. ADGER, W. N., AGRAWALA, S., MIRZA, M. M. Q., CONDE, C., O'BRIEN, K., PULHIN, J., PULWARTY, R., SMIT, B. & TAKAHASHI, K. (eds.) 2007. Assessment of adaptation practices, options, constraints and capacity. In: *Climate Change 2007: Working Group II: Impacts, Adaptation and Vulnerability*. Cambridge, UK: Cambridge University Press.

9. EUROPEAN ENVIRONMENTAL AGENCY. 2013. *Adaptation in Europe – addressing risks and opportunities from climate change in the context of socio-economic developments* [Online]. Available: http://www.eea.europa.eu/publications/adaptation-in-europe (accessed 11 May 2016).

10. JONES, L. 2010. *Overcoming social barriers to climate change adaptation* [Online]. Available: http://www.odi.org.uk/publications/4945-social-barriers-climate-change-adaptation-nepal (accessed 11 May 2016).

11. ADGER, W. N., AGRAWALA, S., MIRZA, M. M. Q., CONDE, C., O'BRIEN, K., PULHIN, J., PULWARTY, R., SMIT, B. & TAKAHASHI, K. (eds.) 2007. Assessment of adaptation practices, options, constraints and capacity. In: *Climate Change 2007: Working Group II: Impacts, Adaptation and Vulnerability*. Cambridge, UK: Cambridge University Press.

12. EUROPEAN ENVIRONMENTAL AGENCY. 2013. *Adaptation in Europe – addressing risks and opportunities from climate change in the context of socio-economic developments* [Online]. Available: http://www.eea.europa.eu/publications/adaptation-in-europe (accessed 11 May 2016).

13. ADGER, W. N., AGRAWALA, S., MIRZA, M. M. Q., CONDE, C., O'BRIEN, K., PULHIN, J., PULWARTY, R., SMIT, B. & TAKAHASHI, K. (eds.) 2007. Assessment of adaptation practices, options, constraints and capacity. In: *Climate Change 2007: Working Group II: Impacts, Adaptation and Vulnerability*. Cambridge, UK: Cambridge University Press.

14. KOLIKOW, S., KRAGT, M. E. & MUGERA, A. W. 2012. *An interdisciplinary framework of limits and barriers to climate change adaptation in agriculture*. Working paper. School of Agricultural and Resource Economics, University of Western Australia, Perth, WA.

15. IPCC. 2014. *AR5 summary for policymakers* [Online]. Available: http://ipcc-wg2.gov/AR5/report/ (accessed 11 May 2016).

16. EUROPEAN ENVIRONMENTAL AGENCY. 2013. *Adaptation in Europe – addressing risks and opportunities from climate change in the context of socio-economic developments* [Online]. Available: http://www.eea.europa.eu/publications/adaptation-in-europe (accessed 11 May 2016).

17. Ibid.

18. HOFFMAN, A. J. 2010. Climate change as a cultural and behavioral issue: addressing barriers and implementing solutions. *Organizational Dynamics*, 39, 295–305.

19. EUROPEAN ENVIRONMENTAL AGENCY. 2013. *Adaptation in Europe – addressing risks and opportunities from climate change in the context of socio-economic developments* [Online]. Available: http://www.eea.europa.eu/publications/adaptation-in-europe (accessed 11 May 2016).

20. NÆSS, P. & HØYER, K. G. 2009. The Emperor's green clothes: growth, decoupling, and capitalism. *Capitalism Nature Socialism*, 20, 74–95.
21. WOLFE, S. & BROOKS, D. B. 2003. Water scarcity: an alternative view and its implications for policy and capacity building. *Natural Resources Forum*, 27, 99–107.
22. Ibid.
23. LIEBERHERR-GARDIOL, F. 2008. Urban sustainability and governance: issues for the twenty-first century. *International Social Science Journal*, 59, 331–342.
24. MILBRATH, L. W. 1995. Psychological, cultural, and informational barriers to sustainability. *Journal of Social Issues*, 51, 101–120.
25. WOLFE, S. & BROOKS, D. B. 2003. Water scarcity: An alternative view and its implications for policy and capacity building. *Natural Resources Forum*, 27, 99–107.
26. HOFFMAN, A. J. 2010. Climate change as a cultural and behavioral issue: addressing barriers and implementing solutions. *Organizational Dynamics*, 39, 295–305.
27. HAUGHTON, G. 1999. Environmental justice and the sustainable city. *Journal of Planning Education and Research*, 18, 233–243.
28. WOLFE, S. & BROOKS, D. B. 2003. Water scarcity: an alternative view and its implications for policy and capacity building. *Natural Resources Forum*, 27, 99–107.
29. STEG, L. & VLEK, C. 2009. Encouraging pro-environmental behaviour: an integrative review and research agenda. *Journal of Environmental Psychology*, 29, 309–317.
30. PAHL-WOSTL, C. 2007. Transitions towards adaptive management of water facing climate and global change. *Water Resources Management*, 21, 49–62.
31. NAJJAR, K. & COLLIER, C. R. 2011. Integrated water resources management: bringing it all together. *Water Resources Impact*, 13, 3–8.
32. PAHL-WOSTL, C. 2007. Transitions towards adaptive management of water facing climate and global change. *Water Resources Management*, 21, 49–62.
33. GLEICK, P. H. 1998. Water in crisis: paths to sustainable water use. *Ecological Applications*, 8, 571–579.
34. SOFOULIS, Z. 2005. Big water, everyday water: a sociotechnical perspective. *Continuum: Journal of Media & Cultural Studies*, 19, 445–463.
35. FARRELLY, M. & BROWN, R. 2011. Rethinking urban water management: experimentation as a way forward? *Global Environmental Change*, 21, 721–732.
36. RICHTER, B. D., ABELL, D., BACHA, E., BRAUMAN, K., CALOS, S., COHN, A., DISLA, C., O'BRIEN, S. F., HODGES, D. & KAISER, S. 2013. Tapped out: how can cities secure their water future? *Water Policy*, 15, 335–363.
37. SMITH, M., DE GROOT, D. & BERGKAMP, G. 2006. *Pay: Establishing Payments for Watershed Services*. Gland: World Conservation Union.
38. WORLD BANK. 2012. Integrated urban water management: a summary note. *Blue Water Green Cities* [Online]. Available: http://siteresources.worldbank.org/INTLAC/Resources/257803-1351801841279/1PrincipalIntegratedUrbanWaterManagement ENG.pdf (accessed 11 May 2016).
39. POLICY RESEARCH INSTITUTE 2005. *Economic Instruments for Water Demand Management in an Integrated Water Resources Management Framework: Synthesis Report*. Policy Research Institute, Government of Canada.
40. UNITED NATIONS. 2013. *UN-water analytical brief on water security and the global water agenda* [Online]. Available: http://www.unwater.org/publications/publications-detail/en/c/197890/ (accessed 3 June 2016).
41. MAHEEPALA, S., BLACKMORE, J., DIAPER, C., MOGLIA, M., SHARMA, A. & KENWAY, S. 2010. Towards the adoption of integrated urban water management approach for planning. *Proceedings of the Water Environment Federation*, 2010, 6734–6753.
42. OECD 2012. *Environmental Outlook to 2050 the Consequences of Inaction: The Consequences of Inaction*. Paris: OECD Publishing.

43. BITHAS, K. 2008. The sustainable residential water use: sustainability, efficiency and social equity. The European experience. *Ecological Economics*, 68, 221–229.

44. ENGEL, K. 2011. *Big cities. Big water. Big challenges: water in an urbanizing world* [Online]. Available: http://www.wwf.se/source.php/1390895/Big%20Cities_Big%20Water_Big%20Challenges_2011.pdf (accessed 11 May 2016).

45. MOLLE, F. 2009. Water and society: new problems faced, new skills needed. *Irrigation and Drainage*, 58, S205–S211.

46. VAN ROON, M. 2007. Water localisation and reclamation: steps towards low impact urban design and development. *Journal of Environmental Management*, 83, 437–447.

47. BITHAS, K. 2008. The sustainable residential water use: sustainability, efficiency and social equity. The European experience. *Ecological Economics*, 68, 221–229.

48. NAJJAR, K. & COLLIER, C. R. 2011. Integrated water resources management: bringing it all together. *Water Resources Impact*, 13, 3–8.

49. UNITED NATIONS. 2013. *UN-water analytical brief on water security and the global water agenda* [Online]. Available: http://www.unwater.org/TFsecurity.html (accessed 3 June 2016).

50. ENGEL, K. 2011. *Big cities. Big water. Big challenges: water in an urbanizing world* [Online]. Available: http://www.wwf.se/source.php/1390895/Big%20Cities_Big%20Water_Big%20Challenges_2011.pdf (accessed 11 May 2016).

51. MOLLE, F. 2009. Water and society: new problems faced, new skills needed. *Irrigation and Drainage*, 58, S205–S211.

52. VAN DER BRUGGE, R. & VAN RAAK, R. 2007. Facing the adaptive management challenge: insights from transition management. *Ecology and Society*, 12, 33.

53. BAHRI, A. 2012. *Integrated urban water management* [Online]. Available: http://www.gwp.org/Global/The%20Challenge/Resource%20material/GWP_TEC16.pdf (accessed 11 May 2016).

54. GLEICK, P. H. 1998. Water in crisis: paths to sustainable water use. *Ecological Applications*, 8, 571–579.

55. WOLFE, S. & BROOKS, D. B. 2003. Water scarcity: an alternative view and its implications for policy and capacity building. *Natural Resources Forum*, 27, 99–107.

56. Ibid.

57. FARRELLY, M. & BROWN, R. 2011. Rethinking urban water management: experimentation as a way forward? *Global Environmental Change*, 21, 721–732.

58. LOUCKS, D. P. 2000. Sustainable water resources management. *Water International*, 25, 3–10.

59. GLOBAL WATER PARTNERSHIP. 2012. *Water demand management (WDM) – the Mediterranean experience*. Technical Focus Paper [Online]. Available: http://www.gwp.org/en/gwp-in-action/News-and-Activities/Global-Water-Partnership-launches-new-publications-at-World-Water-Week-2012/ (accessed 3 June 2016).

60. SOFOULIS, Z. 2005. Big water, everyday water: a sociotechnical perspective. *Continuum: Journal of Media & Cultural Studies*, 19, 445–463.

61. WOLFE, S. & BROOKS, D. B. 2003. Water scarcity: an alternative view and its implications for policy and capacity building. *Natural Resources Forum*, 27, 99–107.

62. NÆSS, P. & HØYER, K. G. 2009. The Emperor's green clothes: growth, decoupling, and capitalism. *Capitalism Nature Socialism*, 20, 74–95.

63. PEARSON, L. J., COGGAN, A., PROCTOR, W. & SMITH, T. F. 2010. A sustainable decision support framework for urban water management. *Water Resources Management*, 24, 363–376.

64. GEORGIA ENVIRONMENTAL PROTECTION DIVISION WATERSHED PROTECTION BRANCH. 2007. *Water conservation education programs EPD Guidance Document* [Online]. Available: http://www1.gadnr.org/cws/Documents/Conservation_Education.pdf (accessed 11 May 2016).

65. PENNSYLVANIA STATE UNIVERSITY. 2010. *Water conservation for communities* [Online]. Available: http://pubs.cas.psu.edu/FreePubs/PDFs/AGRS113.pdf (accessed 11 May 2016).

66. Ibid.

67. BROWNE, A. L., PULLINGER, M., MEDD, W. & ANDERSON, B. 2014. Patterns of practice: a reflection on the development of quantitative/mixed methodologies capturing everyday life related to water consumption in the UK. *International Journal of Social Research Methodology*, 17, 27–43.

68. GEORGIA ENVIRONMENTAL PROTECTION DIVISION WATERSHED PROTECTION BRANCH. 2007. *Water conservation education programs EPD guidance document* [Online]. Available: http://www1.gadnr.org/cws/Documents/Conservation_Education. pdf (accessed 11 May 2016).

69. PENNSYLVANIA STATE UNIVERSITY. 2010. *Water conservation for communities* [Online]. Available: http://pubs.cas.psu.edu/FreePubs/PDFs/AGRS113.pdf (accessed 11 May 2016).

70. GIFFORD, R., KORMOS, C. & MCINTYRE, A. 2011. Behavioral dimensions of climate change: drivers, responses, barriers, and interventions. *Wiley Interdisciplinary Reviews: Climate Change*, 2, 801–827.

71. PENNSYLVANIA STATE UNIVERSITY. 2010. *Water conservation for communities* [Online]. Available: http://pubs.cas.psu.edu/FreePubs/PDFs/AGRS113.pdf (accessed 11 May 2016).

72. GIFFORD, R., KORMOS, C. & MCINTYRE, A. 2011. Behavioral dimensions of climate change: drivers, responses, barriers, and interventions. *Wiley Interdisciplinary Reviews: Climate Change*, 2, 801–827.

73. KEMP, R. & LOORBACH, D. *Governance for sustainability through transition management*. Open Meeting of Human Dimensions of Global Environmental Change Research Community, Montreal, Canada, 2003. 12–30.

74. ELZEN, B. & WIECZOREK, A. 2005. Transitions towards sustainability through system innovation. *Technological Forecasting and Social Change*, 72, 651–661.

75. LOORBACH, D. 2010. Transition management for sustainable development: a prescriptive, complexity-based governance framework. *Governance*, 23, 161–183.

76. KEMP, R. & LOORBACH, D. *Governance for sustainability through transition management*. Open Meeting of Human Dimensions of Global Environmental Change Research Community, Montreal, Canada, 2003. 12–30.

77. DUTCH RESEARCH INSTITUTE FOR TRANSITIONS. 2011. *Urban transition management manual* [Online]. Available: http://www.themusicproject.eu/content/ transitionmanagement (accessed 3 June 2016).

78. GIFFORD, R., KORMOS, C. & MCINTYRE, A. 2011. Behavioral dimensions of climate change: drivers, responses, barriers, and interventions. *Wiley Interdisciplinary Reviews: Climate Change*, 2, 801–827.

79. EUROPEAN COMMISSION. 2013. *EVALSED – the resource for the evaluation of socio-economic development: sourcebook – method and techniques* [Online]. Available: http://ec.europa.eu/regional_policy/en/information/publications/evaluations-guidance-documents/2013/evalsed-the-resource-for-the-evaluation-of-socio-economic-development-sourcebook-method-and-techniques (accessed 3 June 2016).

80. Ibid.

81. AUSTRALIAN GOVERNMENT PRODUCTIVITY COMMISSION. 2012. *Barriers to effective climate change adaptation* [Online]. Available: http://www.pc.gov.au/__data/ assets/pdf_file/0008/119663/climate-change-adaptation.pdf (accessed 3 June 2016).

82. KOLIKOW, S., KRAGT, M. E. & MUGERA, A. W. 2012. *An interdisciplinary framework of limits and barriers to climate change adaptation in agriculture*. Working Paper. School of Agricultural and Resource Economics, University of Western Australia.

83. ADGER, W. N., AGRAWALA, S., MIRZA, M. M. Q., CONDE, C., O'BRIEN, K., PULHIN, J., PULWARTY, R., SMIT, B. & TAKAHASHI, K. (eds.) 2007. Assessment of adaptation practices, options, constraints and capacity. In: *Climate Change 2007: Working Group II: Impacts, Adaptation and Vulnerability*. Cambridge, UK: Cambridge University Press.

84. MOSER, S. C. & EKSTROM, J. A. 2010. A framework to diagnose barriers to climate change adaptation. *Proceedings of the National Academy of Sciences*, 107, 22026–22031.

85. KEMP, R., SCHOT, J. & HOOGMA, R. 1998. Regime shifts to sustainability through processes of niche formation: the approach of strategic niche management. *Technology Analysis & Strategic Management*, 10, 175–198.

86. GIFFORD, R., KORMOS, C. & MCINTYRE, A. 2011. Behavioral dimensions of climate change: drivers, responses, barriers, and interventions. *Wiley Interdisciplinary Reviews: Climate Change*, 2, 801–827.

87. SPENCE, A. & PIDGEON, N. 2009. Psychology, climate change & sustainable behaviour. *Environment. Science and Policy for Sustainable Development*, 51, 8–18.

88. AMERICAN PSYCHOLOGICAL ASSOCIATION. 2009. Psychology and global climate change: addressing a multi-faceted phenomenon and set of challenges. *Task Force on the Interface Between Psychology and Global Climate Change* [Online]. Available: http://www.apa.org/science/about/publications/climate-change.aspx (accessed 3 June 2016).

89. PATCHEN, M. 2010. What shapes public reactions to climate change? Overview of research and policy implications. *Analyses of Social Issues and Public Policy*, 10, 47–68.

90. KOLIKOW, S., KRAGT, M. E. & MUGERA, A. W. 2012. *An interdisciplinary framework of limits and barriers to climate change adaptation in agriculture*. Working Paper. School of Agricultural and Resource Economics, University of Western Australia.

91. MOSER, S. C. & EKSTROM, J. A. 2010. A framework to diagnose barriers to climate change adaptation. *Proceedings of the National Academy of Sciences*, 107, 22026–22031.

92. GLOBAL WATER PARTNERSHIP. 2012. *Water demand management (WDM) – the Mediterranean experience*. Technical Focus Paper [Online]. Available: http://www.gwp.org/en/gwp-in-action/News-and-Activities/Global-Water-Partnership-launches-new-publications-at-World-Water-Week-2012/ (accessed 3 June 2016).

93. GEELS, F. W. 2005. Processes and patterns in transitions and system innovations: refining the co-evolutionary multi-level perspective. *Technological Forecasting and Social Change*, 72, 682.

94. AUSTRALIAN GOVERNMENT PRODUCTIVITY COMMISSION. 2012. *Barriers to effective climate change adaptation* [Online]. Available: http://www.pc.gov.au/__data/assets/pdf_file/0008/119663/climate-change-adaptation.pdf (accessed 3 June 2016).

95. WADE MILLER, G. 2006. Integrated concepts in water reuse: managing global water needs. *Desalination*, 187, 65–75.

96. WARD, F. A., MICHELSEN, A. M. & DEMOUCHE, L. 2007. Barriers to water conservation in the Rio Grande Basin. *JAWRA Journal of the American Water Resources Association*, 43, 237–253.

97. LOCAL GOVERNMENT ASSOCIATION OF NSW. 2012. *Barriers and drivers to sustainability* [Online]. Available: www.lgnsw.org.au/files/imce-uploads/35/barriers-and-drivers-to-sustainability.pdf (accessed 3 June 2016).

98. SEYFANG, G. & SMITH, A. 2007. Grassroots innovations for sustainable development: towards a new research and policy agenda. *Environmental Politics*, 16, 584–603.

99. ADGER, W. N., AGRAWALA, S., MIRZA, M. M. Q., CONDE, C., O'BRIEN, K., PULHIN, J., PULWARTY, R., SMIT, B. & TAKAHASHI, K. (eds.) 2007. Assessment of adaptation practices, options, constraints and capacity. In: *Climate Change 2007: Working Group II: Impacts, Adaptation and Vulnerability*. Cambridge, UK: Cambridge University Press.

100. GIFFORD, R., KORMOS, C. & MCINTYRE, A. 2011. Behavioral dimensions of climate change: drivers, responses, barriers, and interventions. *Wiley Interdisciplinary Reviews: Climate Change*, 2, 801–827.

101. ELZEN, B. & WIECZOREK, A. 2005. Transitions towards sustainability through system innovation. *Technological Forecasting and Social Change*, 72, 651–661.

102. KEMP, R., SCHOT, J. & HOOGMA, R. 1998. Regime shifts to sustainability through processes of niche formation: the approach of strategic niche management. *Technology Analysis & Strategic Management*, 10, 175–198.

103. LOCAL GOVERNMENT ASSOCIATION OF NSW. 2012. *Barriers and drivers to sustainability* [Online]. Available: http://www.lgnsw.org.au/files/imce-uploads/35/barriers-and-drivers-to-sustainability.pdf (accessed 3 June 2016).

104. WARD, S., BARR, S., BUTLER, D. & MEMON, F. A. 2012. Rainwater harvesting in the UK: socio-technical theory and practice. *Technological Forecasting and Social Change*, 79, 1354–1361.

105. SEYFANG, G. & SMITH, A. 2007. Grassroots innovations for sustainable development: towards a new research and policy agenda. *Environmental Politics*, 16, 584–603.

106. AUSTRALIAN GOVERNMENT PRODUCTIVITY COMMISSION. 2012. *Barriers to effective climate change adaptation* [Online]. Available: http://www.pc.gov.au/__data/assets/pdf_file/0008/119663/climate-change-adaptation.pdf (accessed 3 June 2016); SMITH, A., STIRLING, A. & BERKHOUT, F. 2005. The governance of sustainable socio-technical transitions. *Research Policy*, 34, 1491–1510.

107. SMITH, A., STIRLING, A. & BERKHOUT, F. 2005. The governance of sustainable socio-technical transitions. *Research Policy*, 34, 1491–1510.

108. GODWIN, D., PARRY, B., BURRIS, F. & CHAN, S. 2008. *Barriers and opportunities for low impact development* [Online]. Available: http://seagrant.oregonstate.edu/files/sgpubs/onlinepubs/w06002.pdf (accessed 11 May 2016).

109. KOLIKOW, S., KRAGT, M. E. & MUGERA, A. W. 2012. *An interdisciplinary framework of limits and barriers to climate change adaptation in agriculture*. Working Paper. School of Agricultural and Resource Economics, University of Western Australia.

110. MOSER, S. C. & EKSTROM, J. A. 2010. A framework to diagnose barriers to climate change adaptation. *Proceedings of the National Academy of Sciences*, 107, 22026–22031.

111. JONES, L. 2010. *Overcoming social barriers to climate change adaptation.* [Online]. Available: http://www.odi.org.uk/publications/4945-social-barriers-climate-change-adaptation-nepal (accessed 11 May 2016).

112. ADGER, W. N., AGRAWALA, S., MIRZA, M. M. Q., CONDE, C., O'BRIEN, K., PULHIN, J., PULWARTY, R., SMIT, B. & TAKAHASHI, K. (eds.) 2007. Assessment of adaptation practices, options, constraints and capacity. In: *Climate Change 2007: Working Group II: Impacts, Adaptation and Vulnerability*. Cambridge, UK: Cambridge University Press.

113. MULLER, M. 2007. Adapting to climate change: water management for urban resilience. *Environment and Urbanization*, 19, 99–113.

114. FRANTZ, C. M. & MAYER, F. S. 2009. The emergency of climate change: why are we failing to take action? *Analyses of Social Issues and Public Policy*, 9, 205–222.

115. PATCHEN, M. 2010. What shapes public reactions to climate change? Overview of research and policy implications. *Analyses of Social Issues and Public Policy*, 10, 47–68.

116. GEELS, F. W. 2005. Processes and patterns in transitions and system innovations: refining the co-evolutionary multi-level perspective. *Technological Forecasting and Social Change*, 72, 682.

117. GIFFORD, R., KORMOS, C. & MCINTYRE, A. 2011. Behavioral dimensions of climate change: drivers, responses, barriers, and interventions. *Wiley Interdisciplinary Reviews: Climate Change*, 2, 801–827.

118. SMITH, A., STIRLING, A. & BERKHOUT, F. 2005. The governance of sustainable socio-technical transitions. *Research Policy*, 34, 1491–1510.

119. SOFOULIS, Z. 2005. Big water, everyday water: a sociotechnical perspective. *Continuum: Journal of Media & Cultural Studies*, 19, 445–463.

120. BROWN, R. R. & FARRELLY, M. A. 2009. Delivering sustainable urban water management: a review of the hurdles we face. *Water Science & Technology*, 59, 839–846.

121. KEMP, R., SCHOT, J. & HOOGMA, R. 1998. Regime shifts to sustainability through processes of niche formation: the approach of strategic niche management. *Technology Analysis & Strategic Management*, 10, 175–198.

122. ADGER, W. N., AGRAWALA, S., MIRZA, M. M. Q., CONDE, C., O'BRIEN, K., PULHIN, J., PULWARTY, R., SMIT, B. & TAKAHASHI, K. (eds.) 2007. Assessment of adaptation practices, options, constraints and capacity. In: *Climate Change 2007: Working Group II: Impacts, Adaptation and Vulnerability*. Cambridge, UK: Cambridge University Press.

123. KOLIKOW, S., KRAGT, M. E. & MUGERA, A. W. 2012. *An interdisciplinary framework of limits and barriers to climate change adaptation in agriculture*. Working Paper. School of Agricultural and Resource Economics, University of Western Australia.

124. KEMP, R., SCHOT, J. & HOOGMA, R. 1998. Regime shifts to sustainability through processes of niche formation: the approach of strategic niche management. *Technology Analysis & Strategic Management*, 10, 175–198.

125. BROWN, R. R. & FARRELLY, M. A. 2009. Delivering sustainable urban water management: a review of the hurdles we face. *Water Science & Technology*, 59, 839–846.

126. KOLIKOW, S., KRAGT, M. E. & MUGERA, A. W. 2012. *An interdisciplinary framework of limits and barriers to climate change adaptation in agriculture*. Working Paper. School of Agricultural and Resource Economics, University of Western Australia.

127. GLOBAL WATER PARTNERSHIP. 2012. *Water demand management (WDM) – the Mediterranean experience*. Technical Focus Paper [Online]. Available: http://www.gwp.org/en/gwp-in-action/News-and-Activities/Global-Water-Partnership-launches-new-publications-at-World-Water-Week-2012/ (accessed 3 June 2016).

128. MOSER, S. C. & EKSTROM, J. A. 2010. A framework to diagnose barriers to climate change adaptation. *Proceedings of the National Academy of Sciences*, 107, 22026–22031.

129. HOFFMAN, A. J. 2010. Climate change as a cultural and behavioral issue: addressing barriers and implementing solutions. *Organizational Dynamics*, 39, 295–305.

130. SEYFANG, G. & SMITH, A. 2007. Grassroots innovations for sustainable development: towards a new research and policy agenda. *Environmental Politics*, 16, 584–603.

131. BROWN, R. R. & FARRELLY, M. A. 2009. Delivering sustainable urban water management: a review of the hurdles we face. *Water Science & Technology*, 59, 839–846.

132. BÖRZEL, T. & RISSE, T. 2011. From Europeanisation to diffusion: introduction. *West European Politics*, 35, 1–19.

133. SMITH, A., STIRLING, A. & BERKHOUT, F. 2005. The governance of sustainable socio-technical transitions. *Research Policy*, 34, 1491–1510.

134. WALKER, J. L. 1969. The diffusion of innovations among the American states. *The American Political Science Review*, 63, 880–899.

135. GLOBAL WATER PARTNERSHIP. 2012. *Water demand management (WDM) – the Mediterranean experience*. Technical Focus Paper [Online]. Available: http://www.gwp.org/en/gwp-in-action/News-and-Activities/Global-Water-Partnership-launches-new-publications-at-World-Water-Week-2012/ (accessed 3 June 2016).

136. JONES, L. 2010. *Overcoming social barriers to climate change adaptation* [Online]. Available: http://www.odi.org.uk/publications/4945-social-barriers-climate-change-adaptation-nepal (accessed 11 May 2016).

137. PATCHEN, M. 2010. What shapes public reactions to climate change? Overview of research and policy implications. *Analyses of Social Issues and Public Policy*, 10, 47–68.

138. HOFFMAN, A. J. 2010. Climate change as a cultural and behavioral issue: addressing barriers and implementing solutions. *Organizational Dynamics*, 39, 295–305.

139. AUSTRALIAN GOVERNMENT PRODUCTIVITY COMMISSION. 2012. *Barriers to effective climate change adaptation* [Online]. Available: http://www.pc.gov.au/__data/assets/pdf_file/0008/119663/climate-change-adaptation.pdf (accessed 3 June 2016).

140. LOCAL GOVERNMENT ASSOCIATION OF NSW. 2012. *Barriers and drivers to sustainability* [Online]. Available: http://www.lgnsw.org.au/files/imce-uploads/35/barriers-and-drivers-to-sustainability.pdf (accessed 3 June 2016).

141. GERO, A., KURUPPU, N. & MUKHEIBIR, P. 2012. *Cross-Scale Barriers to Climate Change Adaptation in Local Government, Australia.* Sydney, NSW: UTS: Institute for Sustainable Futures, University of Technology Sydney.

142. GODWIN, D., PARRY, B., BURRIS, F. & CHAN, S. 2008. *Barriers and opportunities for low impact development* [Online]. Available: http://seagrant.oregonstate.edu/files/sgpubs/onlinepubs/w06002.pdf (accessed 11 May 2016).

143. FURLONG, K., Cook, C. & Bakker, K. 2008. *Good governance for water conservation: a primer* [Online]. Available: http://watergovernance.ca/wp-content/uploads/2010/02/Good-Governance-Primer.pdf (accessed 11 May 2016).

144. LOCAL GOVERNMENT ASSOCIATION OF NSW. 2012. *Barriers and drivers to sustainability* [Online]. Available: http://www.lgnsw.org.au/files/imce-uploads/35/barriers-and-drivers-to-sustainability.pdf (accessed 3 June 2016).

145. AUSTRALIAN GOVERNMENT PRODUCTIVITY COMMISSION. 2012. *Barriers to effective climate change adaptation* [Online]. Available: http://www.pc.gov.au/__data/assets/pdf_file/0008/119663/climate-change-adaptation.pdf (accessed 3 June 2016).

146. MOSER, S. C. & EKSTROM, J. A. 2010. A framework to diagnose barriers to climate change adaptation. *Proceedings of the National Academy of Sciences*, 107, 22026–22031.

147. HOFFMAN, A. J. 2010. Climate change as a cultural and behavioral issue: addressing barriers and implementing solutions. *Organizational Dynamics*, 39, 295–305.

148. PATCHEN, M. 2010. What shapes public reactions to climate change? Overview of research and policy implications. *Analyses of Social Issues and Public Policy*, 10, 47–68.

149. GEELS, F. W. 2005. Processes and patterns in transitions and system innovations: refining the co-evolutionary multi-level perspective. *Technological Forecasting and Social Change*, 72, 682.

150. ECONOMIC AND SOCIAL RESEARCH COUNCIL. 2008. *Behavioural change and water efficiency* [Online]. Available: http://webarchive.nationalarchives.gov.uk/20080821115857/http://esrc.ac.uk/ESRCInfoCentre/about/CI/events/esrcseminar/BehaviouralChangeandWaterEfficiency.aspx?ComponentId=25751&SourcePageId=6066 (accessed 11 May 2016).

151. SMITH, A., STIRLING, A. & BERKHOUT, F. 2005. The governance of sustainable socio-technical transitions. *Research Policy*, 34, 1491–1510.

152. AUSTRALIAN GOVERNMENT PRODUCTIVITY COMMISSION. 2012. *Barriers to effective climate change adaptation* [Online]. Available: http://www.pc.gov.au/__data/assets/pdf_file/0008/119663/climate-change-adaptation.pdf (accessed 3 June 2016).

153. PELLING, M. 2010. *Adaptation to Climate Change: From Resilience to Transformation.* London: Taylor & Francis.

154. ELZEN, B. & WIECZOREK, A. 2005. Transitions towards sustainability through system innovation. *Technological Forecasting and Social Change*, 72, 651–661.

155. KOLIKOW, S., KRAGT, M. E. & MUGERA, A. W. 2012. *An interdisciplinary framework of limits and barriers to climate change adaptation in agriculture*. Working Paper. School of Agricultural and Resource Economics, University of Western Australia.

156. PELLING, M. 2010. *Adaptation to Climate Change: From Resilience to Transformation*. New York: Taylor & Francis.

157. KEMP, R., SCHOT, J. & HOOGMA, R. 1998. Regime shifts to sustainability through processes of niche formation: the approach of strategic niche management. *Technology Analysis & Strategic Management*, 10, 175–198.

158. PELLING, M. 2010. *Adaptation to Climate Change: From Resilience to Transformation*. New York: Taylor & Francis.

159. ELZEN, B. & WIECZOREK, A. 2005. Transitions towards sustainability through system innovation. *Technological Forecasting and Social Change*, 72, 651–661.

160. AUSTRALIAN GOVERNMENT PRODUCTIVITY COMMISSION. 2012. *Barriers to effective climate change adaptation* [Online]. Available: http://www.pc.gov.au/__data/assets/pdf_file/0008/119663/climate-change-adaptation.pdf (accessed 3 June 2016).

161. KEMP, R., SCHOT, J. & HOOGMA, R. 1998. Regime shifts to sustainability through processes of niche formation: the approach of strategic niche management. *Technology Analysis & Strategic Management*, 10, 175–198.

162. LORENZONI, I., NICHOLSON-COLE, S. & WHITMARSH, L. 2007. Barriers perceived to engaging with climate change among the UK public and their policy implications. *Global Environmental Change*, 17, 445–459.

163. WENDT, A. 1999. *Social Theory of International Politics*. Cambridge, UK: Cambridge University Press.

164. SEYFANG, G. & SMITH, A. 2007. Grassroots innovations for sustainable development: towards a new research and policy agenda. *Environmental Politics*, 16, 584–603.

165. Ibid.

166. ADGER, W. N., AGRAWALA, S., MIRZA, M. M. Q., CONDE, C., O'BRIEN, K., PULHIN, J., PULWARTY, R., SMIT, B. & TAKAHASHI, K. (eds.) 2007. Assessment of adaptation practices, options, constraints and capacity. In: *Climate Change 2007: Working Group II: Impacts, Adaptation and Vulnerability*. Cambridge, UK: Cambridge University Press.

167. ROGERS, E. M. 2003. *Diffusion of Innovations*, 5th Edition. New York: Free Press.

168. MOSER, S. C. & EKSTROM, J. A. 2010. A framework to diagnose barriers to climate change adaptation. *Proceedings of the National Academy of Sciences*, 107, 22026–22031.

169. AUSTRALIAN GOVERNMENT PRODUCTIVITY COMMISSION. 2012. *Barriers to effective climate change adaptation* [Online]. Available: http://www.pc.gov.au/__data/assets/pdf_file/0008/119663/climate-change-adaptation.pdf (accessed 3 June 2016).

170. GERO, A., KURUPPU, N. & MUKHEIBIR, P. 2012. *Cross-Scale Barriers to Climate Change Adaptation in Local Government, Australia*. UTS: Institute for Sustainable Futures, University of Technology Sydney.

171. LANE, H., KROGH, C. & O'FARRELL, L. 2007. Identifying social attitudes and barriers to water conservation – a community water survey. In: AUSTRALIA, E. (ed.) *Rainwater and Urban Design*. Barton, A.C.T.: Engineers Australia.

172. PATCHEN, M. 2010. What shapes public reactions to climate change? Overview of research and policy implications. *Analyses of Social Issues and Public Policy*, 10, 47–68.

173. KOLIKOW, S., KRAGT, M. E. & MUGERA, A. W. 2012. *An interdisciplinary framework of limits and barriers to climate change adaptation in agriculture*. Working Paper. School of Agricultural and Resource Economics, University of Western Australia.

174. MOSER, S. C. & EKSTROM, J. A. 2010. A framework to diagnose barriers to climate change adaptation. *Proceedings of the National Academy of Sciences*, 107, 22026–22031.

175. GERO, A., KURUPPU, N. & MUKHEIBIR, P. 2012. *Cross-Scale Barriers to Climate Change Adaptation in Local Government, Australia.* UTS: Institute for Sustainable Futures, University of Technology Sydney.

176. KOLLMUSS, A. & AGYEMAN, J. 2002. Mind the gap: why do people act environmentally and what are the barriers to pro-environmental behavior? *Environmental Education Research*, 8, 239–260.

177. AMERICAN PSYCHOLOGICAL ASSOCIATION. 2009. Psychology and global climate change: addressing a multi-faceted phenomenon and set of challenges. *Task Force on the Interface Between Psychology and Global Climate Change* [Online]. Available: http://www.apa.org/science/about/publications/climate-change.aspx (accessed 3 June 2016).

178. SMITH, G. 2005. Green citizenship and the social economy. *Environmental Politics*, 14, 273–289.

179. PIKE, C., DOPPELT, B. & HERR, M. 2010. *Climate communications and behavior change: a guide for practitioners.* The Climate Leadership Initiative, Eugene, OR [Online]. Available: http://www.thesocialcapitalproject.org/The-Social-Capital-Project/pubs/climate-communications-and-behavior-change (accessed 3 June 2016).

180. BRECHIN, S. R. & BHANDARI, M. 2011. Perceptions of climate change worldwide. *Wiley Interdisciplinary Reviews: Climate Change*, 2, 871–885.

181. GIFFORD, R., KORMOS, C. & MCINTYRE, A. 2011. Behavioral dimensions of climate change: drivers, responses, barriers, and interventions. *Wiley Interdisciplinary Reviews: Climate Change*, 2, 801–827.

182. MCKENZIE-MOHR, D. 2000. New ways to promote proenvironmental behavior: promoting sustainable behavior: an introduction to community-based social marketing. *Journal of Social Issues*, 56, 543–554.

183. Ibid.

184. GODWIN, D., PARRY, B., BURRIS, F. & CHAN, S. 2008. *Barriers and opportunities for low impact development* [Online]. Available: http://seagrant.oregonstate.edu/files/sgpubs/onlinepubs/w06002.pdf (accessed 11 May 2016).

185. AUSTRALIAN GOVERNMENT PRODUCTIVITY COMMISSION. 2012. *Barriers to effective climate change adaptation* [Online]. Available: http://www.pc.gov.au/__data/assets/pdf_file/0008/119663/climate-change-adaptation.pdf (accessed 3 June 2016).

186. AMERICAN PSYCHOLOGICAL ASSOCIATION. 2009. Psychology and global climate change: addressing a multi-faceted phenomenon and set of challenges. *Task Force on the Interface Between Psychology and Global Climate Change* [Online]. Available: http://www.apa.org/science/about/publications/climate-change.aspx (accessed 3 June 2016).

187. MILBRATH, L. W. 1995. Psychological, cultural, and informational barriers to sustainability. *Journal of Social Issues*, 51, 101–120.

188. POLICY RESEARCH INSTITUTE 2005. *Economic Instruments for Water Demand Management in an Integrated Water Resources Management Framework: Synthesis Report.* Policy Research Institute, Government of Canada.

189. MCKENZIE-MOHR, D. 2000. New ways to promote proenvironmental behavior: promoting sustainable behavior: an introduction to community-based social marketing. *Journal of Social Issues*, 56, 543–554.

190. MILBRATH, L. W. 1995. Psychological, cultural, and informational barriers to sustainability. *Journal of Social Issues*, 51, 101–120.

191. RESER, J. P. & BENTRUPPERBÄUMER, J. M. 2005. What and where are environmental values? Assessing the impacts of current diversity of use of 'environmental' and 'World Heritage' values. *Journal of Environmental Psychology*, 25, 125–146.

192. PIKE, C., DOPPELT, B. & HERR, M. 2010. *Climate communications and behavior change: a guide for practitioners*. The Climate Leadership Initiative [Online]. Available: http://www.thesocialcapitalproject.org/The-Social-Capital-Project/pubs/climate-communications-and-behavior-change (accessed 3 June 2016).

193. PATCHEN, M. 2010. What shapes public reactions to climate change? Overview of research and policy implications. *Analyses of Social Issues and Public Policy*, 10, 47–68.

194. GIFFORD, R., KORMOS, C. & MCINTYRE, A. 2011. Behavioral dimensions of climate change: drivers, responses, barriers, and interventions. *Wiley Interdisciplinary Reviews: Climate Change*, 2, 801–827.

195. SCHULTZ, P. 2011. Conservation means behavior. *Conservation Biology*, 25, 1080–1083.

196. BALMFORD, A. & COWLING, R. M. 2006. Fusion or failure? The future of conservation biology. *Conservation Biology*, 20, 692–695.

197. FRANTZ, C. M. & MAYER, F. S. 2009. The emergency of climate change: why are we failing to take action? *Analyses of Social Issues and Public Policy*, 9, 205–222.

198. GODWIN, D., PARRY, B., BURRIS, F. & CHAN, S. 2008. *Barriers and opportunities for low impact development* [Online]. Available: http://seagrant.oregonstate.edu/files/sgpubs/onlinepubs/w06002.pdf (accessed 11 May 2016).

199. AUSTRALIAN GOVERNMENT PRODUCTIVITY COMMISSION. 2012. *Barriers to effective climate change adaptation* [Online]. Available: http://www.pc.gov.au/__data/assets/pdf_file/0008/119663/climate-change-adaptation.pdf (accessed 3 June 2016).

200. AMERICAN PSYCHOLOGICAL ASSOCIATION. 2009. Psychology and global climate change: addressing a multi-faceted phenomenon and set of challenges. *Task Force on the Interface Between Psychology and Global Climate Change* [Online]. Available: http://www.apa.org/science/about/publications/climate-change.aspx (accessed 3 June 2016).

201. KOLLMUSS, A. & AGYEMAN, J. 2002. Mind the gap: why do people act environmentally and what are the barriers to pro-environmental behavior? *Environmental Education Research*, 8, 239–260.

202. GIFFORD, R., KORMOS, C. & MCINTYRE, A. 2011. Behavioral dimensions of climate change: drivers, responses, barriers, and interventions. *Wiley Interdisciplinary Reviews: Climate Change*, 2, 801–827.

203. MILBRATH, L. W. 1995. Psychological, cultural, and informational barriers to sustainability. *Journal of Social Issues*, 51, 101–120.

204. PATCHEN, M. 2010. What shapes public reactions to climate change? Overview of research and policy implications. *Analyses of Social Issues and Public Policy*, 10, 47–68.

205. FRANTZ, C. M. & MAYER, F. S. 2009. The emergency of climate change: why are we failing to take action? *Analyses of Social Issues and Public Policy*, 9, 205–222.

206. SPENCE, A. & PIDGEON, N. 2009. Psychology, climate change & sustainable behaviour. *Environment: Science and Policy for Sustainable Development*, 51, 8–18.

207. PATCHEN, M. 2010. What shapes public reactions to climate change? Overview of research and policy implications. *Analyses of Social Issues and Public Policy*, 10, 47–68.

208. FRANTZ, C. M. & MAYER, F. S. 2009. The emergency of climate change: why are we failing to take action? *Analyses of Social Issues and Public Policy*, 9, 205–222.

209. SPENCE, A. & PIDGEON, N. 2009. Psychology, climate change & sustainable behaviour. *Environment: Science and Policy for Sustainable Development*, 51, 8–18.

210. PIKE, C., DOPPELT, B. & HERR, M. 2010. *Climate communications and behavior change: a guide for practitioners*. The Climate Leadership Initiative [Online]. Available: http://www.thesocialcapitalproject.org/The-Social-Capital-Project/pubs/climate-communications-and-behavior-change (accessed 3 June 2016).

211. Ibid.

212. PATCHEN, M. 2010. What shapes public reactions to climate change? Overview of research and policy implications. *Analyses of Social Issues and Public Policy*, 10, 47–68.

213. GIFFORD, R., KORMOS, C. & MCINTYRE, A. 2011. Behavioral dimensions of climate change: drivers, responses, barriers, and interventions. *Wiley Interdisciplinary Reviews: Climate Change*, 2, 801–827.

214. SPENCE, A. & PIDGEON, N. 2009. Psychology, climate change & sustainable behaviour. *Environment: Science and Policy for Sustainable Development*, 51, 8–18.

215. SCHULTZ, P. 2011. Conservation means behavior. *Conservation Biology*, 25, 1080–1083.

216. FRANTZ, C. M. & MAYER, F. S. 2009. The emergency of climate change: why are we failing to take action? *Analyses of Social Issues and Public Policy*, 9, 205–222.

217. AMERICAN PSYCHOLOGICAL ASSOCIATION. 2009. Psychology and global climate change: addressing a multi-faceted phenomenon and set of challenges. *Task Force on the Interface Between Psychology and Global Climate Change* [Online]. Available: http://www.apa.org/science/about/publications/climate-change.aspx (accessed 3 June 2016).

218. BRECHIN, S. R. & BHANDARI, M. 2011. Perceptions of climate change worldwide. *Wiley Interdisciplinary Reviews: Climate Change*, 2, 871–885.

219. PIKE, C., DOPPELT, B. & HERR, M. 2010. *Climate communications and behavior change: a guide for practitioners.* The Climate Leadership Initiative [Online]. Available: http://www.thesocialcapitalproject.org/The-Social-Capital-Project/pubs/climate-communications-and-behavior-change (accessed 3 June 2016).

220. GIFFORD, R., KORMOS, C. & MCINTYRE, A. 2011. Behavioral dimensions of climate change: drivers, responses, barriers, and interventions. *Wiley Interdisciplinary Reviews: Climate Change*, 2, 801–827.

221. AMERICAN PSYCHOLOGICAL ASSOCIATION. 2009. Psychology and global climate change: addressing a multi-faceted phenomenon and set of challenges. *Task Force on the Interface Between Psychology and Global Climate Change* [Online]. Available: http://www.apa.org/science/about/publications/climate-change.aspx (accessed 3 June 2016).

222. GIFFORD, R., KORMOS, C. & MCINTYRE, A. 2011. Behavioral dimensions of climate change: drivers, responses, barriers, and interventions. *Wiley Interdisciplinary Reviews: Climate Change*, 2, 801–827.

223. EUROPEAN COMMISSION. 2012. Green behaviour. *Future Brief* [Online]. Available: http://ec.europa.eu/environment/integration/research/newsalert/pdf/FB4.pdf (accessed 11 May 2016).

224. SCHULTZ, P. 2011. Conservation means behavior. *Conservation Biology*, 25, 1080–1083.

225. PIKE, C., DOPPELT, B. & HERR, M. 2010. *Climate communications and behavior change: a guide for practitioners.* The Climate Leadership Initiative [Online]. Available: http://www.thesocialcapitalproject.org/The-Social-Capital-Project/pubs/climate-communications-and-behavior-change (accessed 3 June 2016).

226. PATCHEN, M. 2010. What shapes public reactions to climate change? Overview of research and policy implications. *Analyses of Social Issues and Public Policy*, 10, 47–68.

227. KAPLAN, S. 2000. New ways to promote proenvironmental behavior: human nature and environmentally responsible behavior. *Journal of Social Issues*, 56, 491–508.

228. MILBRATH, L. W. 1995. Psychological, cultural, and informational barriers to sustainability. *Journal of Social Issues*, 51, 101–120.

229. HOFFMAN, A. J. 2010. Climate change as a cultural and behavioral issue: addressing barriers and implementing solutions. *Organizational Dynamics*, 39, 295–305.

230. FRANTZ, C. M. & MAYER, F. S. 2009. The emergency of climate change: why are we failing to take action? *Analyses of Social Issues and Public Policy*, 9, 205–222.

231. KEMP, R., SCHOT, J. & HOOGMA, R. 1998. Regime shifts to sustainability through processes of niche formation: the approach of strategic niche management. *Technology Analysis & Strategic Management*, 10, 175–198.

232. GIFFORD, R., KORMOS, C. & MCINTYRE, A. 2011. Behavioral dimensions of climate change: drivers, responses, barriers, and interventions. *Wiley Interdisciplinary Reviews: Climate Change*, 2, 801–827.

233. PATCHEN, M. 2010. What shapes public reactions to climate change? Overview of research and policy implications. *Analyses of Social Issues and Public Policy*, 10, 47–68.

234. DOLNICAR, S. & HURLIMANN, A. 2010. Australians' water conservation behaviours and attitudes. *Australian Journal of Water Resources*, 14, 43–53.

235. HOFFMAN, A. J. 2010. Climate change as a cultural and behavioral issue: addressing barriers and implementing solutions. *Organizational Dynamics*, 39, 295–305.

236. PATCHEN, M. 2010. What shapes public reactions to climate change? Overview of research and policy implications. *Analyses of Social Issues and Public Policy*, 10, 47–68.

237. KAPLAN, S. 2000. New ways to promote proenvironmental behavior: human nature and environmentally responsible behavior. *Journal of Social Issues*, 56, 491–508.

238. AMERICAN PSYCHOLOGICAL ASSOCIATION. 2009. Psychology and global climate change: addressing a multi-faceted phenomenon and set of challenges. *Task Force on the Interface Between Psychology and Global Climate Change* [Online]. Available: http://www.apa.org/science/about/publications/climate-change.aspx (accessed 3 June 2016).

239. GIFFORD, R., KORMOS, C. & MCINTYRE, A. 2011. Behavioral dimensions of climate change: drivers, responses, barriers, and interventions. *Wiley Interdisciplinary Reviews: Climate Change*, 2, 801–827.

240. PIKE, C., DOPPELT, B. & HERR, M. 2010. *Climate communications and behavior change: a guide for practitioners*. The Climate Leadership Initiative [Online]. Available: http://www.thesocialcapitalproject.org/The-Social-Capital-Project/pubs/climate-communications-and-behavior-change (accessed 3 June 2016).

241. Ibid.

242. BRECHIN, S. R. & BHANDARI, M. 2011. Perceptions of climate change worldwide. *Wiley Interdisciplinary Reviews: Climate Change*, 2, 871–885.

243. FRANTZ, C. M. & MAYER, F. S. 2009. The emergency of climate change: why are we failing to take action? *Analyses of Social Issues and Public Policy*, 9, 205–222.

244. PATCHEN, M. 2010. What shapes public reactions to climate change? Overview of research and policy implications. *Analyses of Social Issues and Public Policy*, 10, 47–68.

245. HOFFMAN, A. J. 2010. Climate change as a cultural and behavioral issue: addressing barriers and implementing solutions. *Organizational Dynamics*, 39, 295–305.

246. PATCHEN, M. 2010. What shapes public reactions to climate change? Overview of research and policy implications. *Analyses of Social Issues and Public Policy*, 10, 47–68.

247. AMERICAN PSYCHOLOGICAL ASSOCIATION. 2009. Psychology and global climate change: addressing a multi-faceted phenomenon and set of challenges. *Task Force on the Interface Between Psychology and Global Climate Change* [Online]. Available: http://www.apa.org/science/about/publications/climate-change.aspx (accessed 3 June 2016).

248. SCHULTZ, P. 2011. Conservation means behavior. *Conservation Biology*, 25, 1080–1083.

249. ADGER, W. N., AGRAWALA, S., MIRZA, M. M. Q., CONDE, C., O'BRIEN, K., PULHIN, J., PULWARTY, R., SMIT, B. & TAKAHASHI, K. (eds.) 2007. Assessment of adaptation practices, options, constraints and capacity. In: *Climate Change 2007: Working Group II: Impacts, Adaptation and Vulnerability*. Cambridge, UK: Cambridge University Press.

250. SCHULTZ, P. 2011. Conservation means behavior. *Conservation Biology*, 25, 1080–1083.

251. PIKE, C., DOPPELT, B. & HERR, M. 2010. *Climate communications and behavior change: a guide for practitioners*. The Climate Leadership Initiative [Online]. Available:http://www.thesocialcapitalproject.org/The-Social-Capital-Project/pubs/climate-communications-and-behavior-change (accessed 3 June 2016).

252. PATCHEN, M. 2010. What shapes public reactions to climate change? Overview of research and policy implications. *Analyses of Social Issues and Public Policy*, 10, 47–68.

253. BROWNE, A. L., PULLINGER, M., MEDD, W. & ANDERSON, B. 2014. Patterns of practice: a reflection on the development of quantitative/mixed methodologies capturing everyday life related to water consumption in the UK. *International Journal of Social Research Methodology*, 17, 27–43.

7 Amsterdam transitioning towards urban water security

Introduction

Amsterdam's transition towards urban water security prioritises the development of an efficient mains system with low leakage, efficient supply of high-quality water at a low price and the recovery of raw material and energy from water and waste. This case study analyses how Amsterdam's water utility uses a portfolio of demand management tools to modify the attitudes and behaviour of water users to achieve urban water security.

The case study first provides a brief company background of Amsterdam's water utility, along with an overview of the city's water supply and water consumption levels before discussing the city's strategic vision for achieving urban water security. The case study will then analyse the various demand management tools used by the city's water utility in an attempt to achieve urban water security before discussing the numerous barriers identified by the utility in achieving further urban water security in Amsterdam.

7.1 Brief company background

Waternet is the only water company in the Netherlands that is dedicated to the entire cycle, from providing drinking water to collecting and treating storm and wastewater. Waternet produces drinking water and treats wastewater. Waternet maintains water levels and keeps surface water clean. This is done on behalf of the Regional Public Water Authority of Amstel, Gooi and Vecht and the City of Amsterdam.

Urban Water Security, First Edition. Robert C. Brears.
© 2017 John Wiley & Sons, Ltd. Published 2017 by John Wiley & Sons, Ltd.

7.2 Water supply and water consumption

Waternet provides drinking water to 1.2 million people in Amsterdam and the wider Amsterdam Metropolitan Area (over 800 000 reside in Amsterdam City and the remainder in the Metropolitan Area). Waternet's philosophy is that the utility provides its customers with tap water that is completely reliable and does not require any chemical treatment to ensure its quality. Waternet provides drinking water that is sourced from both groundwater (60 percent) and surface water (40 percent). It is naturally treated before being supplied to customers. The natural treatment steps are as follows:

- *Natural lake*: Beside the Loenderveense Lake is the Waterworks Lake, where Amsterdam's drinking water is made.
- *Purification*: Groundwater is supplemented with water from the Amsterdam–Rhine Canal. Ferric chloride removes suspended contaminated particles before the water is added to the Waterworks Lake. The water stays in the Waterworks Lake for around hundred days where it undergoes a natural self-cleaning process that degrades ammonia, organic substances and bacteria. By adding hydrochloric acid, the acidity of the water (pH) is controlled.
- *Rapid sand filtration*: The water then filters through bins filled with layers of sand and gravel removing pollutants.
- *Transportation*: A 10 kilometre transmission line brings the water to the last treatment plant where the taste and colour of the water are improved.
- *Post-treatment*: By adding ozone, pesticides and pathogens are broken down. Also a large portion of the calcium is removed. The water is then filtered again. The water first passes through a container of activated charcoal before it is filtered through fine sand. After this process the water is ready for consumption.
- *Warehousing and distribution*: Waternet stores the water in drinking water reservoirs from which it is pumped through a pipeline system, 2000 kilometres long, to the utility's customers.

Amsterdam's per capita daily consumption has decreased from 139.1 litres in 2009 to around 134 litres per day in 2014 (Table 7.1). In Amsterdam around 60 percent of customers are domestic and 40 percent are non-domestic: Amsterdam has a low industrial base compared to Rotterdam, which has a large harbour with many industries.

Table 7.1 Water consumption in Amsterdam

Year	Daily per capita consumption (litres)
2009	139.1
2010	138.6
2011	134.8
2012	133.5
2013	133.3
2014	133.5–134 (projected)

7.3 Strategic vision: Amsterdam's Definitely Sustainable 2011–2014

The 'Amsterdam Definitely Sustainable 2011–2014' plan outlines a comprehensive programme for achieving sustainability based on four pillars: climate and energy (using renewable energy and the efficient use of fossil fuels to lower carbon emissions within the city), mobility and air quality (Amsterdam will be a reachable city with a sustainable transport system), sustainable innovative economy (national and international companies will choose Amsterdam to conduct sustainability-related business) and materials and consumers (Amsterdam is a liveable city where its citizens and companies use raw materials in an effective way). The city of Amsterdam believes that its primary responsibility is to develop and implement specific urban solutions in order to realise its transition towards sustainability, especially because urban areas, including Amsterdam, are positioned to lead the greening of the global economy through improvements in areas including water and waste systems.[1]

The effective use of raw materials is one of the key aspects of the Amsterdam Definitely Sustainable plan, where effective cycles contribute towards the efficient use of materials and resources and therefore reduce the city's ecological footprint. Regarding water in the circular economy, the city's water cycle prioritises, first, an efficient main system with low leakage, second, the supply of high-quality water at a low price and, third, the recovery of raw material and energy from water and waste, which forms the backbone of Amsterdam becoming a more sustainable city.

7.4 Drivers of water security

The drivers of Waternet's strategic vision for achieving urban water security include corporate rebranding, protecting good quality raw water and human health, political and economic factors, carbon neutrality, population growth and climate change (summarised in Figure 7.1).

7.4.1 Corporate rebranding

The campaign towards water conservation started in the 1990s when drinking water companies began to term themselves environmentally sustainable companies. This led to water companies having two objectives: first, reusing waste materials and, second, making customers behave in a way that reduces water consumption. Today, water companies in the Netherlands, including Waternet, try to reuse all the waste materials produced; however, there is less focus on reducing water consumption as it is already low at around 130 litres per person per day, and the Netherlands is not considered a water-stressed country.

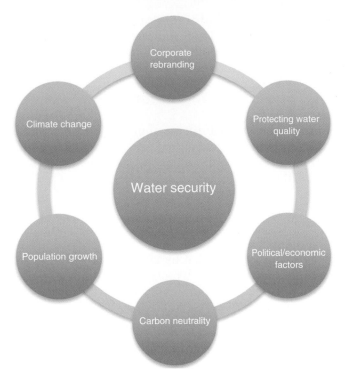

Figure 7.1 Drivers of water security in Amsterdam.

7.4.2   Protecting good quality raw water and human health

In the Netherlands water quantity is not a problem but water quality is – not the quality of drinking water but the quality of raw water that is polluted due to the Netherlands being a very dense country with a lot of agricultural production: the country has problems with pesticides in the raw water sources. In addition, there are problems with pharmaceutical products in the raw water sources. However, the public demands drinking water to be of spring-like quality and unchlorinated: in the 1970s, Dr. Joop Rook, a Dutch water scientist, discovered the presence of trihalomethanes in chlorinated drinking water – carcinogenic compounds that are by-products of disinfection, formed as a result of chlorination of natural organic matter. This discovery revolutionised the Dutch drinking water industry with chlorination being banned and alternative treatment processes developed. As such all water utilities in the Netherlands have done their best to ensure that drinking water is free of chlorine. Therefore, the focus is on how to protect the quality of water sources. There are two possible approaches for Waternet in ensuring quality of drinking water: either protect drinking water sources upstream or introduce additional drinking water treatment systems downstream. If additional treatment systems are implemented, the question Waternet asks is whether this is sustainable. First, it means society, according to Waternet, is accepting of pollution. Second, Waternet would have to introduce additional treatment steps that would require additional energy, which could possibly increase carbon emissions. As a result, the water sector in the Netherlands, including Waternet, is focused on the protection of drinking water sources.

7.4.3   Political and economic

A driver of sustainability is the mayor of Amsterdam working with a public relations company to promote Amsterdam as a modern, sustainable city in order to attract young people, scientists and new businesses and drive green economic growth of the city.

7.4.4   Carbon neutrality

Amsterdam is aiming that by 2025 the city will have reduced its greenhouse gas emissions by 40 percent, which is a reduction of 3 100 000 tonnes CO_2-eq per year. Meanwhile, Waternet aims to be climate neutral in 2020, requiring reductions in greenhouse gas emissions of 53 000 ton CO_2-eq per year.

7.4.5   Population growth

Between 2009 and 2011, the growth rate of Amsterdam, along with Rotterdam and The Hague, was 3 percent compared to 1 percent for the Netherlands as a whole. Over that period, Amsterdam's population grew by 25 000 people, from around 755 000 to 780 000 inhabitants. In 2013 Amsterdam's population increased to over 800 000.[2]

7.4.6   Climate change

Climate change is likely to affect water resources in the Netherlands. While the Rhine's average discharge will increase in winter, by up to 12 percent, it will decrease in summer, by up to 23 percent. If there is no change in airflow patterns over Western Europe, average summer precipitation will increase by 3–6 percent. However, if easterly winds prevail, rainfall could decrease by 10–19 percent. Nonetheless, the chance of extreme drought is greater because at higher temperatures evaporation will exceed rainfall levels.[3] With rising sea levels and lower river discharges in the summer months, there will be increased salinisation of drinking water: salt water will penetrate further inland reducing the number of days freshwater inlet points can be used. At the same time, the amount of water that can be taken from the main system to combat internal salinisation will decrease while demand increases.[4] As such, climate change is likely to expose Amsterdam to lower levels of surface water from the Rhine, which the utility uses to 'top' up the groundwater it uses. In addition, salinisation of groundwater supplies will likely diminish the number of wells available for drinking water, reducing the overall supply of water for the city.

7.5   Regulatory and technological demand management tools to achieve urban water security

7.5.1   Drinking water and wastewater tariffs

Waternet's drinking water tariffs are based on full cost recovery. If a customer has a water meter, their water bill will consist of a variable rate of EUR 1.24 per cubic metre of water consumed, a fixed rate of EUR 42.15 per year and a tax on drinking water (Table 7.2); the total amount of tax depends on the number of cubic metres of water the customer consumes. In addition there is a 6 percent value added tax (VAT).

There are two tariffs Waternet charges for wastewater: one for sewage transport (fixed tariff) and the other for wastewater treatment in which the size of the tariff depends on the size of the household – if a person lives alone, they pay for one person equivalent (fixed tariff) of approximately EUR 50 a year; households with two or more (it does not matter if the household contains two, three, four, five, six or more) pay for the equivalent of three people per year (~EUR 150).

Billing and low-income support

Waternet's customers receive a water bill once per quarter that is based on an estimate of how much water the customer consumed; the estimate is based on the water consumption in previous years. For customers struggling to pay their water bill, Waternet tries to make appointments with these customers on how they can pay, for example, providing customers the ability to make delayed payments. Nonetheless, customers will not be provided water for free or at a lower price.

7.5.2   Metering

Prior to 1997 Amsterdam's water users were not metered as the city had a different tariff system: customers had to pay according to the amount of rooms and taps in their house. Since 1997 it was decided by the town council that water meters

Table 7.2 Water tax for metered customers

Water consumption (cubic metre)	Rate (per cubic metre)
0–300	€0.33
300–50 000	€0.40
50 000–250 000	€0.36
250 000–1 250 000	€0.26
>1 250 000	€0.05

should be installed for two reasons. First, customers have to pay exactly for what they use. Second, when customers know exactly what they use, they want to use water more wisely. The result is less water use and more sustainable use of water. Today around 70 percent of houses have a water meter with the utility aiming for universal metering (individual metering for all connections including sub-metering in apartment buildings). To achieve this, meters are installed in houses that are renovated. However, it will be another 10–15 years before Waternet achieves universal metering. Meanwhile, non-domestic customers are fully metered.

Smart metering

There has been a large discussion in the Netherlands about the introduction of smart meters. Until now smart meters have not been introduced in Amsterdam or any other city in the Netherlands because it lacks a positive business case. It only has a positive business case if smart meters are connected with the energy sector. However, the water sector does not want to be dependent on the energy sector, so customers have traditional, manual-read water meters. With smart meters the water companies would be able to read meters much more efficiently, and it would enable customers to directly see how much water they use. Furthermore, smart meters would enable the whole metering process to be more customer friendly. Currently, customers have to read their own meter once a year. In addition there is a lack of clear access to meters with many customers' meters located in cellars.

7.5.3 Reducing unaccounted-for water

Waternet has an extremely low unaccounted-for water (UFW) rate of 2–3 percent. To maintain low leakage rates, Waternet uses a top-down method for computing leakage that is common in the Netherlands. This method involves the water balance where the difference between the system's input volume and the revenue water is calculated. The difference is referred to as non-revenue water which consists of water used by the company, for example, for cleaning and disinfecting of mains; unbilled consumption, for example, firefighting; temporary non-metred consumption and illegal consumption; meter inaccuracies; administrative errors and leakage. All components excluding leakage are estimated. Leakage is then calculated by subtracting the other components from non-revenue water. This is in comparison with the bottom-up method that calculates leakage by analysing flow at night when water usage is low.

If measurements indicate a high level of leakage in a certain area, the exact location can be found using noise loggers and leak noise correlators. However, these two pieces of equipment have the disadvantage in that the maximum distance for recording the sounds of a leak depends on factors including the pipe material. For metallic pipes, distances of several hundred metres can be recorded compared to cement pipes of around a hundred metres. With plastic pipes the maximum length is restricted to tens of metres. Low leakage rates in

Amsterdam and the Netherlands as a whole are due to a range of factors including:

- The majority of the water distribution networks are laid after 1950 with the most common material being PVC, followed by asbestos cement and cast iron.
- The majority of mains are in the footpath with loose block paving on the surface; therefore, leaks are readily visible.
- Soil types vary between sand, clay and peat, and the soil contains little stones. Pipes are laid on levelled sand beds and backfilled with excavated soil.
- Because the Netherlands is relatively flat, pressures within the distribution system range from 250 to 500 kilopascal.
- Burst rates of mains range from 0.05 to 0.15 burst per kilometre per year, including bursts by third parties.
- Most service pipes are made of polyethylene or copper. All lead services have been replaced and the service pipes contain minimal joints.
- No active leak control is carried out, only passive leak control, that is, responding to reported bursts.
- Most areas have universal metering.[5]

7.5.4 Protecting the quality of source water

In order to protect the quality of source and lower water treatment costs, Waternet has a voluntary programme with farmers to encourage them to use fewer pesticides or no pesticides at all. The programme has two components: first, farmers are educated on how pesticides can contaminate sources of water leading to unhealthy drinking water for all users including farmers themselves – the aim is to encourage farmers to use less pesticides in their operations. Second, the programme encourages farmers to create zones within their agricultural areas where they do not use any pesticides at all. To incentivise farmers to do so, Waternet provides farmers with financial incentives to compensate them for not using pesticide products (also on a European level within the Water Framework Directive, there is the ability to use financial incentives for not using some certain pesticides).

7.5.5 Reducing energy costs in wastewater treatment

Waternet's main wastewater treatment plant 'Amsterdam West' is located beside a waste-to-energy plant operated by AEB Waste to Energy Company. The close proximity enables an exchange of energy flows between the two plants with large environmental benefits: Amsterdam West produces 25 000 cubic metres per day of biogas and 100 000 tons of sewage sludge per year for burning at the waste-to-energy plant. The energy produced in the waste-to-energy plant is then used to power the Amsterdam West treatment plant. In total, the integration of the two plants produces 20 000 megawatt hours per year of electricity and 50 000 gigajoule per year of heat, saving 1.8 million cubic metres per year of natural gas, resulting in avoided greenhouse gas emissions of 3200 tons per year.[6]

7.5.6   Alternative water supplies

Rainwater harvesting and greywater reuse have been a topic in the Netherlands since the 1990s. There were some experiments conducted with greywater, but it led to problems with cross-connections between the greywater and drinking water supply. This led to an entire housing estate receiving water of a lower quality and people becoming ill. Therefore, the government abandoned the use of greywater systems in the Netherlands and legislated against its use. Regarding rainwater harvesting, customers can use rainwater for toilet flushing but to do so requires special permission from the health inspectorate. To gain permission, the system has to be proved that it is safe, in particular the customer must have a risk management plan based on a risk assessment that all steps have been taken to avoid cross-connections. As a result of this bureaucratic paperwork, there are hardly any rainwater harvesting systems in the Netherlands.

7.6   Communication and information demand management tools to achieve urban water security

7.6.1   School programmes: Sight visits and education programmes

For schools, Waternet has a special programme where children visit Waternet's plants and nature conservation areas and learn about water-related issues. In addition, Waternet has developed numerous education programmes that target school children in promoting awareness of water and sustainability, which are listed in Table 7.3.

Table 7.3 Waternet education programmes

Education programme	Description
Green linked	Anyone of any age can search for information on nature, sustainability and education
Droppie water	Designed for elementary school students to learn about clean water, sanitation, flooding and the work of water boards
Water wise	People can find everything about water management in the Netherlands. In TV clips, animations, games, etc., people can learn about how water is treated and also about flood protection
Video: clean, safe and adequate water	The animation shows what the duties of the Dutch water boards are and how they provide clean, adequate and safe water
Water chats	Water chats is a game about creating safe levees and providing sufficient clean water
Sieb the sheep	Sheep Sieb takes you into the world of rivers. To avoid flooding the rivers need more space. But how do you do that? Look quickly at the movies!
Education ships	Children and young people can learn about water on the water itself
Guest lectures on water	Experts come in to tell the class about water in the Netherlands

7.6.2   Public education: Determining the message

There are two ways Waternet determines which topics the utility should focus on in water-awareness campaigns: first, Waternet uses feedback from customers (via letters, emails, etc.) to determine future campaign topics. Waternet's marketing department then conducts, on an infrequent basis, campaigns to raise awareness on these water-related issues, for example, sewage (informing people what should/should not be poured down their drains – grease from cooking often clogs the sewage pipes) and the benefits of softened water (people can use less washing powder). Second, the Association of Dutch Water Companies often plans and coordinates campaigns for all Dutch water companies to run at the same time: as there are only 10 water companies in the Netherlands, coordination becomes easy.

7.6.3   Promotion of water-efficient devices

Waternet provides information to customers on the availability of water-saving devices such as low-flow showerheads and more efficient washing machines on the market, etc. At times, Waternet gives away tap faucets for free at public events; however, it is not Waternet's policy to do this on a large scale.

7.6.4   Billing inserts

Billing inserts are frequently used as a tool of communication. When Waternet has certain campaigns, for example, promoting the benefits of softened water, the utility inserts a brochure in with the customer's bill.

7.6.5   Promoting water-efficient technologies

Waternet conducts R&D projects within Amsterdam and uses the home market to test water technologies. The next step, which is driven by the mayor, is to promote these water technologies to other cities. However, Waternet is a public company and cannot risk public funds to market these new water technologies. Therefore, Waternet is establishing a new entity called Clean Capital that will enter into strategic partnerships with private sector companies to market these technologies.

7.6.6   Non-domestic water efficiency advice

Large-scale water users can seek advice from Waternet on how they can save water and be more efficient in their operations.

7.7 Case study SWOT analysis

7.7.1 Strengths

Socially, Waternet is driven by the desire to be seen by society as a sustainable, responsible water company. Economically, Amsterdam is promoting itself as a sustainable city to encourage sustainability-related companies to locate there.

Regarding the environment, Waternet is protecting the quality of raw water as the Netherlands is a densely populated country with significant amounts of groundwater pollution. Waternet is also aiming to protect the environment by reducing its greenhouse gas emissions and recycling waste from wastewater into energy.

Due to the flat topography and use of modern pipes, Waternet's UFW rate is very low, around 2–3 percent. This means the water utility can supply drinking water services to its customers efficiently, reducing energy costs and carbon emissions from having to provide excess water.

While Waternet does not promote water conservation as such, the utility promotes the wise use of water as its conservation leads to energy savings from treating less wastewater, which in turn lowers carbon emissions. Furthermore, by treating less wastewater it means Waternet's wastewater treatment plants will have additional capacity for treating more storm water during heavy storm events that are likely to become more frequent with climate change.

Waternet has an in-depth children's education programme to ensure that children of all ages understand the need to be wise with water. The programme includes video games, classroom visits and educational excursions.

7.7.2 Weaknesses

Waternet's pricing model does not promote water conservation for heavy users of water – the tax on water actually decreases as consumption levels increase. This harms the 'image' of Amsterdam as a city that promotes the wise use of water, which is the basis of the city attracting R&D investments in water technologies and engineering solutions. This also hinders the utility's ability to promote the sustainable and wise use of water by domestic users, as there is the potential for them to see 'big business' as not doing enough due to water being cheap.

Waternet's domestic customers are not fully metered, hindering the ability of the utility to ensure customers are using water wisely. Waternet needs to become proactive in promoting water meters by, for example, providing subsidies for the installation of water meters in homes.

Waternet does not use financial incentives to promote water-efficient technologies such as low-flow showerheads or fix leaking toilets.

The government has legislated against the use of greywater systems in the Netherlands, despite these systems reducing water consumption levels and energy costs in providing non-potable water. Regarding rainwater harvesting, customers can use rainwater for toilet flushing but to do so requires special permission from the health inspectorate, which requires a lengthy risk assessment to ensure there are no cross-connections. As a result of this bureaucratic burden, rainwater harvesting systems are seldom implemented in Amsterdam or the rest of the Netherlands.

7.7.3   Opportunities

Regarding reducing costs in the wastewater treatment system, Waternet could introduce a wastewater tariff that has a fixed and variable component with the variable component providing an incentive for customers to further reduce the amount of water requiring treatment. Overall, this would reduce energy costs and increase the capacity of wastewater treatment plants to handle large volumes of water during storm events.

To further encourage customers to have meters installed, the utility could promote the amount of energy or carbon emissions customers would save through reductions in water usage. There is the potential for Waternet to introduce AMR metering on a limited scale, for businesses, schools, universities, public organisations, etc. and promote to the wider population their successes in using water wisely (reducing water consumption levels).

To further promote using water wisely, Waternet should implement a monthly billing system: there is currently a disconnect in time between customer behaviour and billing information. If the water and energy sector could combine to implement an AMR metering system throughout Amsterdam, Waternet could move towards monthly billing based on actual reads, rather than estimates of water use based on consumption levels from the previous year. A monthly billing system would also enable Waternet to develop a deeper relationship with customers on their need to use water wisely and protect the wastewater treatment system.

While Waternet provides customers with information on water-saving devices on the market and at times distributes these devices for free at public events, the utility does not provide any economic incentives (subsidies/rebates) for customers to purchase the devices. If customers were encouraged to do so, it could lower energy costs of treating wastewater (and carbon emissions). In addition, the utility could provide subsidies for the replacement of older water-using household appliances with newer more efficient ones, reducing water and energy use. Finally, Waternet could provide subsidies/rebates for the installation of rainwater harvesting systems that could flush toilets or replenish groundwater supplies, which in turn would lower the costs of providing potable water and treating wastewater (energy) and reduce carbon emissions.

Currently, non-domestic customers can approach the utility for advice on water conservation. To enhance water awareness, Waternet can develop a simple water audit programme for non-domestic users, from which the utility could monitor and promote businesses that have reduced consumption levels significantly. Case studies could be developed for other businesses to emulate.

In addition, Waternet could promote water audits for members of specific associations such as the local hotel association: important given Amsterdam is a major tourist destination.

Waternet should use demographic data to better identify customer segments for subsidy/rebate programmes. For instance, the utility provides additional time to customers having difficulty in paying their bills; therefore, Waternet could have social welfare organisations provide low-income families with subsidy/ rebate coupons for the installation of water-saving devices throughout their homes. Waternet could even promote subsidies/rebates for the installation of low-flow toilets. Waternet can use demographic data to be more specific in distributing water-saving devices; rather than distributing devices to people at public events which is a self-selected audience, the utility can organise competitions for certain groups in which winners receive the devices while at the same time enabling others of the same demographic group to emulate those winners. In addition, Waternet can use demographic data to better tailor the messages in their billing inserts.

To truly close the loop and decouple water supply from energy use, carbon emissions and population growth, Amsterdam needs to promote the use of rainwater and greywater systems as alternative viable sources of supply, enabling the city to match demand for water with varying levels of quality (as currently potable water is used to flush toilets). To gain public acceptance of alternative supplies, the city and Waternet could develop stringent standards for alternative systems that ensure there is no backflow (contamination of drinking water supplies from untreated wastewater). The city and its utility could then encourage investments in reuse systems that incorporate these standards. In addition, Waternet could investigate the feasibility of establishing at the neighbourhood level a decentralised system that separates water of differing quality for different uses such as watering gardens. Overall, with Amsterdam already testing water technologies and exporting solutions for sea-level rise and flood management, the city can further enhance its status as a hydro-hub by promoting the development and testing of alternative water supply systems that will solve global water scarcity issues.

7.7.4   Threats

In the Netherlands, climate change is likely to affect the availability of water resources, with the Rhine's average discharge increasing in winter and decreasing in summer. In addition, the probability of extreme drought will increase over time. Meanwhile, rising sea levels and lower river discharges in the summer months will increase the likelihood of salinisation of drinking water. At the same time the amount of water that can be taken from the main system to combat internal salinisation will decrease while demand increases.

In Amsterdam, climate change is likely to expose the city to lower levels of surface water from the Rhine, which Waternet uses to 'top' up the groundwater the city uses. In addition, salinisation of groundwater supplies will likely diminish the number of wells available for drinking water, reducing the overall supply of water for the city. At the same time, the population of Amsterdam is rapidly growing, increasing demand for water resources that may be scarce in the coming decades.

7.8 Transitioning towards urban water security summary

Waternet uses a portfolio of demand management tools to achieve urban water security (Table 7.4). However, there are numerous barriers identified by the utility in achieving further urban water security in Amsterdam (Table 7.5).

Table 7.4 Demand management tools to achieve urban water security

Diffusion mechanisms	Tools	Description
Manipulation of utility calculations	Pricing of drinking water	Waternet's tariffs for water are based on recovering the full economic costs of providing potable water
	Pricing of wastewater	Waternet charges customers for transporting and treating sewage
	Partnerships with private companies	Clean Capital partners with private sector companies to market water-efficient technologies
	Source protection incentives	Farmers at times are compensated for not using pesticide products
Legal and physical coercion	Metering	In 1997, Amsterdam began to meter its domestic customers. Currently, 70 percent of houses have meters. Waternet is aiming to have universal metering within the next 10–15 years. All non-domestic customers are fully metered
	UFW	Waternet has an extremely low UFW rate of 2–3 percent
	Source protection	Waternet has a voluntary programme that encourages farmers to limit or refrain from using pesticides
Socialisation	Promotion of water-efficient devices	Tap faucets are given away for free at public events
Persuasion	Education in schools	School visits and classroom materials
		Online games and cartoons
	Public education	Information on water-saving devices
		Customer feedback guides awareness and Association of Dutch Water Companies guide public awareness topics
		Billing inserts are used to communicate with customers
		Advice on water conservation for large water users on request
Emulation	Reducing energy costs and carbon emissions	Energy from biogas and sewage sludge is used at the wastewater treatment plant

Table 7.5 Barriers to further urban water security

Barrier	Description
Economic/ technological	Smart meters lack a positive business case unless connected with the energy sector. However, the water sector does not want to be dependent on the energy sector
Economic	With the current revenue model, it is difficult to promote water conservation because revenues will decrease
Regulatory/ technological	A greywater experiment led to people becoming ill from cross-connections. As such, the government has legislated against its use
	Rainwater-harvesting systems require significant amounts of paperwork before installation
Regulatory	For large consumers of water, the tax on water decreases after a certain amount
Institutional	There is no regular programme for large-scale users on water conservation
Political	Waternet does not use financial incentives to promote water-efficient technologies
Demographics	Waternet has not conducted targeted water conservation campaigns based on certain demographic groups

Notes

1. CITY OF AMSTERDAM. 2011. A green metropole: Amsterdam definitely sustainable [Online]. Available: https://www.amsterdam.nl/publish/pages/.../a_green_metropole_def.pdf (accessed 10 May 2016).
2. STATISTICS NETHERLANDS. 2014. Population growth concentrates in 30 largest municipalities [Online]. Available: http://www.cbs.nl/en-GB/menu/themas/bevolking/publicaties/artikelen/archief/2014/2014-4056-wm.htm (accessed 10 May 2016).
3. RIJKSWATERSTAAT: MINISTRY OF INFRASTRUCTURE AND THE ENVIRONMENT. 2011. Water management in the Netherlands. Available: http://www.rijkswaterstaat.nl/en/images/Water%20Management%20in%20the%20Netherlands_tcm224-303503.pdf (accessed 10 May 2016).
4. Ibid.
5. BEUKEN, R. H. S., LAVOOIJ, C. S. W., BOSCH, A. & SCHAAP, P. G. Low leakage in the Netherlands confirmed. Eighth Annual Water Distribution Systems Analysis Symposium 2006, Cincinnati, OH, USA, 2006.
6. VAN DER HOEK, J. P., STRUKER, A. & DANSCHUTTER, J. E. M. 2013. Amsterdam as a Sustainable European Metropolis: Integration of Water, Energy and Material Flows. *International Water Week*, Amsterdam, The Netherlands, 4–6 November 2013.

8 Berlin transitioning towards urban water security

Introduction

Berlin's transition towards urban water security focuses on educating the public on how to use water in the right way as well as reducing energy usage and carbon emissions in the city's water and wastewater system. This case study analyses how Berlin's water utility uses a portfolio of demand management tools to modify the attitudes and behaviour of water users to achieve urban water security.

The case study first provides a brief company background of Berlin's water utility, along with an overview of the city's water supply and water consumption levels before discussing the city's strategic vision for achieving urban water security. The case study will then analyse the various demand management tools used by the city's water utility in an attempt to achieve urban water security before discussing the numerous barriers identified by the utility in achieving further urban water security in Berlin.

8.1 Brief company background

Berlin's water utility, Berliner Wasserbetriebe (BWB), is more than 150 years old. After the reunification of Germany, the eastern and western parts of the company came together to form BWB. In 1999, the City of Berlin decided to privatise BWB

Urban Water Security, First Edition. Robert C. Brears.
© 2017 John Wiley & Sons, Ltd. Published 2017 by John Wiley & Sons, Ltd.

by selling 49.9 percent of the shares. However, today BWB is now 100 percent owned by the City of Berlin with the Senate repurchasing in 2012 the outstanding shares from RWE, an electricity company in Germany, and Aeolia, a large French company. BWB's turnover is around EUR 1.1 billion, and the utility has around 4000 employees that provide the city with drinking water and wastewater treatment services.

8.2   Water supply and water consumption

BWB supplies water to around 3.5 million inhabitants of Berlin. BWB has around 700 deep wells, reaching depths of 170 metres, which provide water to the city's nine waterworks, from which the water is treated before being distributed to consumers through almost 8000 kilometres of pipeline network. All of the company's waterworks are located inside the city area, with one exception, which is the Stolpe waterworks located on the outskirts of the city. Overall, BWB prefers to sell around 190 million cubic metres of water per annum for the whole city.

BWB does not treat its water chemically to ensure clean healthy water; instead the utility provides spring-quality groundwater to its customers via a natural treatment process that removes ligand (iron) and mangal from the groundwater – the reason is when the two come into contact with oxygen it settles in the pipeline system eventually blocking the pipes. Furthermore, there are two aesthetic reasons for its removal: first, white shirts turn grey, and second, the taste is similar to blood. To remove iron and mangal, the water passes through sand filters before being stored in large storage tanks, from which the water is pumped directly to households. To ensure the quality of drinking water, BWB monitors the quality of water every day. In total, BWB takes around 45 000 samples per year from its network to ensure it is providing healthy, good quality water.

The composition of BWB's customer base is 70 percent domestic and 30 percent non-domestic. Regarding water consumption levels, in 1989, water consumption in West Berlin was around 150 litres per capita per day while in East Berlin it was around 300 litres. Since the reunification of Germany and the merging of the eastern and western parts of the city's water company, water consumption for all users (domestic and non-domestic) in Berlin has decreased by 45 percent. Today, Berlin's per capita daily water consumption for all users is around 120 litres. Meanwhile, consumption rates for domestic users in Berlin, after unification, decreased by 18 percent over the period 1992–2009: from 138 litres per capita per day to 113 litres per capita per day. Overall, water consumption in Berlin is decreasing by 1.5 percent per annum. BWB states that there are three main reasons for decreasing consumption: introduction of new water-efficient technologies that consume less water, rising awareness about water scarcity globally and high virtual water consumption in Germany and people being convinced that saving water is consistent with being eco-friendly.

8.3   Strategic vision: Using water wisely

In the aftermath of World War II, Berlin was divided into separate administrative sectors: a western sector controlled by the United States, Britain and France and an eastern sector controlled by the Soviet Union. With Germany divided into the Federal Republic of Germany (West Germany) and the German Democratic Republic (East Germany), West Berlin became part of West Germany despite being surrounded by East Germany. In 1961, the Berlin Wall was constructed, further isolating West Berlin physically. As a result, West Berlin was reliant on water resources from within its own administrative boundaries, forcing the city to adopt a closed water cycle approach.[1] Specifically, West Berlin had to extract water from within the city's boundaries, reduce the quantity of water used for various purposes, ensure groundwater withdrawal was proportionate to recharge rates, ensure the city's water bodies were protected from pollution as strictly as possible, use treated wastewater to increase the flow rate of water bodies and rely on stormwater retention to complement other limited resources.[2]

Since the fall of the Berlin Wall, Berlin has continued to promote the sustainable use of the city's own water resources. In 2000, Berlin's Senate passed new legislation requiring all water used in Berlin to be abstracted from within the city's boundaries and promote a more responsible and sustainable use of its water resources.[3] Today, BWB has ceased promoting water conservation as such because of rising groundwater levels from lower consumption levels: when consumption decreased significantly (less groundwater was being withdrawn to meet supply), groundwater levels increased so much that in many parts of Berlin today, houses/buildings cannot be constructed as the water table is too high – therefore, it is counterproductive for the utility to promote water conservation. Instead, BWB is focused on educating the public on how to use water in the right way. In addition, BWB aims to reduce energy usage and carbon emissions in its water and wastewater system.

8.3.1   Berlin Water Act

Berlin's Water Act states that as a component of the ecosystem, water resources are to be managed in a way that serves the general good, and their uses by individuals does not impact the harmony of this. In addition, it should be ensured 'that avoidable adverse effects on their ecological functions and on the terrestrial ecosystems and wetlands directly dependent on them do not occur with respect to their water balance, and that through this sustainable development is guaranteed overall'.[4]

8.4   Drivers of water security

The drivers of BWB's strategic vision for achieving urban water security include the need to protect water supply from wastewater contamination, reduce energy costs and carbon emissions and adapt to climate change (summarised in Figure 8.1).

Figure 8.1 Drivers of water security in Berlin.

8.4.1 Protecting water supply from wastewater contamination

As Berlin's drinking water wells are located on the riverbanks of the city's rivers, it is important that wastewater is treated to a very high standard; if not the city's water supply may become contaminated. One of the challenges to ensuring clean drinking water is that with an ageing population more medication is entering wastewater. In the inner part of Berlin, the city's rainwater and sewage water is collected in a combined system with nearly 10 000 kilometres of sewage pipelines. As Berlin is not topographically flat, the pipelines transport the combined rain and sewage water to collection points located in nine deep depressions through-out the city, from which 150 pumping stations move the collected wastewater through 1000 kilometres of pressure pipelines to wastewater treatment plants located on the outskirts of the city – with one exception, the Ruhleben plant, which is located inside the city. At these wastewater treatment plants, screens take out solid materials and grit from the streets – normally 50–60 tonnes of waste and 30 tonnes of grit are removed per week. The next step involves primary cleaning where anything that has settled at the bottom of the tanks is removed. Afterwards the wastewater is processed through activated sludge tanks which remove, through biological processes, nitrates and phosphorus. The final process involves the clean water being processed through a finer clarifier before it is pumped onto the surface water of Berlin. Meanwhile, on the outskirts of Berlin, the utility operates a separated system where rainwater and sewage water is collected separately with collected rainwater along with water collected from the highways used to irrigate fields, which then replenishes the city's groundwater supplies.

8.4.2 Reducing energy costs and carbon emissions

Using renewable energy in the water and wastewater plants is very important for BWB as it decreases energy costs and reduces carbon emissions: in 2008, BWB signed a commitment that the utility would reduce its CO_2-emissions by 35 000 tonnes per year until 2010.

8.4.3 Climate change impacting water availability

In the eastern regions of Germany, climate change models predict a 20 percent increase in rainfall over the winter period, which will be balanced out by a decrease in rainfall over summer. In addition, warmer temperatures will lower snowmelt reducing surface water levels and groundwater recharge levels. Berlin is vulnerable to heavy sporadic precipitation, which can lower the availability of good quality water due to contamination of groundwater supplies and increase the threat of flooding. As Berlin and the surrounding Brandenburg region will likely experience low annual rainfall and the soil is sandy (so it retains little water), the region as a whole is vulnerable to reduced groundwater levels and droughts.

8.5 Regulatory and technological demand management tools to achieve urban water security

8.5.1 Tariff for drinking water and wastewater

In Berlin, the tariff for drinking water is composed of a fixed and variable component. The fixed amount depends on the size of the water meter (Table 8.1); for instance, if a household's water meter size is 2.5 cubic metres and the house uses 100 cubic metres of water per annum, the household will pay EUR 0.045 per day net or EUR 0.048 gross per day. The variable component for water users is a volumetric tariff of EUR 2.169 per cubic metre.

For wastewater (sewage and rainwater), the tariff is composed of a variable and fixed rate. The volumetric tariff for sewage water is EUR 1.464 per cubic metre. For rainwater, customers pay an annual fixed amount based on the property's sealed surface area: EUR 1.82 per square metre (sqm) of sealed surface multiplied by the size of the sealed surface. For example, if a customer has a property of 900 sqm and 100 sqm of that is 'sealed' surface (house, garage and driveway), the customer will pay the sum of, in this case, EUR 1.82 multiplied by 100 sqm for the rainwater: EUR 182 per annum.

Setting of the tariffs

The procedure to fix these tariffs is based on a guideline that lists costs to be included in the pricing of water: material costs, staff costs, rental costs and operating costs. Interest costs from BWB's loans are included in the tariff as imputed costs, meaning BWB will pay a defined interest rate on capital employed.

Table 8.1 Fixed tariff for domestic customers

Water meter size	Consumption	Fixed tariff (net) euros	Fixed tariff (gross) euros
2.5	0–100	0.045	0.048
2.5	101–200	0.060	0.064
2.5	201–400	0.099	0.106
2.5	401–1000	0.198	0.212
2.5	Above 1001	0.300	0.321

Every 2 years the tariffs are adjusted to meet real costs. If customers are overcharged they are refunded.

Before BWB can publish the tariff rates, the tariff calculations are reviewed by the City of Berlin. As an additional oversight, large accounting firms such as PwC and KPMG review the tariff calculations. Nonetheless, in 2012, the Berlin's court system found BWB's tariffs to be too high and ordered the utility to repay customers EUR 60 million each year for the next three financial years (2012–2015). Furthermore, the courts will investigate the history of BWB's tariffs: if the tariffs are found to be too high, the court will order BWB to repay another EUR 60 million to customers for the last three financial years (2009–2012). If so, BWB will need to take out a loan to pay the costs as the utility's revenue is not large enough

8.5.2   Metering

BWB provides water to around 3.5 million inhabitants of Berlin. Despite this, BWB has only around 250 000 customers, each fully metered. Private houses have their own water meters in Berlin; however, whether an apartment in an apartment complex has an individual water meter (submeter) is dependent on who owns the apartment: inhabitants who own their apartment have a submeter and therefore are customers of BWB. For rented flats it is not the tenants who have a direct connection to BWB but the landlord, which is usually a large company owning the entire apartment complex: these types of buildings have one water meter for the entire apartment block with the water bill divided among the tenants by the landlord, who is the customer of BWB. Another issue of not metering every user of water directly is ensuring customer satisfaction, as the utility cannot directly contact the other 3.25 million users of its services.

Automatic meter readers

BWB is providing customers with new automatic meter readers (AMRs). The AMRs transfer back to BWB's headquarters customer's water usage data. However, only around 6 percent of BWB's customers have AMRs as they are expensive. For normal meters, BWB has to change the meters every 5 years and while doing so the meters are read. In between these 5 years, customers provide their meter figure via telephone or the Internet directly to BWB's customer service office, which then sends out a bill. Regarding smart meters, there is a smart meter pilot project in Potsdamer Platz; however, BWB avoids installing them on a regular basis as they are easy to manipulate.

8.5.3   Reducing unaccounted-for water

BWB's supply area is divided into five water districts. Each district has a service network operating centre, whose staff is responsible for the servicing and maintenance of the mains and pipes. Each year they carry out around 5300 repair jobs, of which around 2000 are due to pipe bursts in supply and building connections. This alone requires around 21 000 road excavations per annum. In order to reduce leakage and pipe bursts, BWB employees annually check more than 68 000 fittings in the pipe network. In addition, around 2 kilometres of pipes are cleaned annually and lined with cement mortar to improve their flow rate. Water mains are also checked systematically for leaks every 4 years. As a result Berlin has a very low unaccounted-for water (UFW) rate of less than 4 percent.[5]

Each year BWB invests EUR 250–270 million in the city's water and sewage distribution networks as well as its waterworks and sewage water plants. Before reunification, East Berlin had a different strategy from West Berlin in how it invested in the water distribution system. In the eastern part, leakages were repaired, but there was little investment in the water distribution network as a whole. Meanwhile in the western part, the water utility invested in the network by conducting preventative maintenance: upgrading of pipes before leakages occur. The difference could be seen at the beginning of reunification with the western side of the city having very low levels of UFW compared to the eastern side's UFW rate of about 25 percent.

Following reunification, BWB invested in the network on the eastern side by implementing its successful preventive maintenance programme, reducing UWF in the eastern part to around 4–5 percent. Through continued investment, the utility expects to see that by 2018 UFW on the former eastern side will be the same level as the western part. In addition to the normal increase in pipeline bursts during winter time, there was a spike in pipeline bursts during the construction of Potsdamer Platz and the renewal of eastern parts of the city with construction workers often accidentally bursting water pipelines.

8.5.4   Source protection: Reducing treatment costs

As BWB's wells are located inside the city, the utility has to protect its groundwater supplies from contamination, which increases treatment costs (energy) and carbon emissions. To do so, almost one third of Berlin is designated as lying in water protection areas. Inside these areas, certain activities are forbidden, for example, there should be no airport or industry inside these areas. Nonetheless, compromises are made such as allowing the part of Tegel Airport that lies slightly within the protected area to remain as it is. Specifically there are three zones of protection for groundwater wells in Berlin (summarised in Figure 8.2):

- *Zone III*: This area extends approximately 2.5 kilometres around the wells. Within this perimeter anything that can contaminate the reserves or impair the taste of the groundwater is strictly prohibited. This includes discharging wastewater, cooling water and condensation or even rainwater (except stormwater runoff from roofs). Housing complexes and industrial commercial facilities are not permitted to be built unless they are connected to the local public sewer network. Parking, washing or repairing motor vehicles is not permitted on unpaved soil.
- *Zone II*: This area extends at least 100 metres from the wells and serves to protect the groundwater hygienically, particularly from pathogens. In this zone the continuous presence of people and animals, or the removal or destruction of the upper soil layer, is strictly prohibited. This includes the construction and renovation of buildings, excavations and transport of liquids hazardous to water as well as the transportation of rubble and waste. Furthermore, it is prohibited to keep animals for commercial purposes and to use natural fertilisers and pesticides in this area.

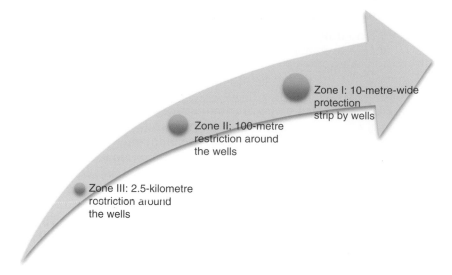

Figure 8.2 Source protection of groundwater supplies.

- *Zone I*: This area is a 10 metre-wide strip on both sides of a row of wells. Any activity involving the upper layer of soil in the immediate vicinity of groundwater extraction facilities or any activity that can risk contamination is strictly prohibited.

8.5.5 Alternative water supplies

BWB does not utilise greywater in Berlin to reduce the amount of potable water entering the sewer system. This is despite potable water requiring energy to treat and pump water. There is a potential of cross-connections where contaminated greywater enters the drinking water system which will not only affect the customer's health but may also lead to the contamination of the entire water treatment plants. Therefore, BWB avoids having these systems installed.

BWB does not promote rainwater harvesting; however there are a couple of experiments in the city, most notably Potsdamer Platz, where rainwater is collected from 19 buildings over an estimated area of 32 000 square metres and then stored in a 35 000 cubic metre rainwater basement tank, from which the collected water is used for flushing toilets, watering of gardens and replenishment of an artificial pond. With a separated system, rainwater can be used to replenish the groundwater levels; however, this becomes difficult with many rainwater harvesting systems on the market in Germany made out of copper, potentially contaminating groundwater resources.

8.5.6 Reducing energy costs

Since 2009, BWB operates the largest photovoltaic plant in Berlin at the utility's Tegel waterworks plant. The solar energy generated is enough to provide drinking water supply to 8500 Berliners saving 157 tonnes of CO_2 per annum. Furthering that commitment BWB in 2012 installed three wind turbines at its wastewater treatment plant in Schönerlinde. While the cost of installing the turbines was EUR 11 million each, the three wind turbines combined produce 80–90 percent of total energy required to run the plant, saving BWB significant energy costs. In addition, BWB uses a fuel-efficient vehicle fleet to reduce its carbon emissions.

8.5.7   Reducing treatment costs: Separate systems

On the outskirts of Berlin, BWB operates a separated wastewater system (sewage is collected separately to rainwater). By operating this type of system, the costs of treating wastewater is reduced in two ways: first, operational costs of treating wastewater is lowered as the treatment plants only have to treat sewage water. Second, infrastructural costs of building new treatment plants are lower for separate systems as they process less wastewater – if treatment plants are built to handle increased precipitation during storm events as a result of climate change, they need to be very large, for example, BWB's Ruhleben plant, which is in the middle of the city, has a dry weather capacity of 300 000 cubic metres of water per day; however its wet weather capacity is 650 000 cubic metres of water per day. Therefore, any treatment plant that handles both sewage and rainwater has to be more than double the size of a plant that only treats sewage water.

8.5.8   Water-efficient technologies

After reunification, water consumption in East Berlin decreased due to, first, its inhabitants purchasing more water-efficient household appliances and, second, old apartments being renovated and water-efficient technologies installed. This decrease in consumption is still ongoing: each year, water consumption in Berlin decreases by 1.5 percent because of renovations and newer water-efficient technologies.

8.6   Communication and information demand management tools to achieve urban water security

8.6.1   Water awareness in the past

In the past, when Berlin's water consumption rates began to exceed the 330 million cubic metres of groundwater available to the city, BWB initiated a water-saving campaign in which the utility would visit kindergartens and schools to raise awareness on the need to conserve water. Regarding the general public, BWB conducted water-saving campaigns through newspaper advertisements. Furthermore, the utility had a travelling exhibition that attended public events to, first, increase the visibility of the utility and, second, educate the public on how to save water. In particular, the exhibition would demonstrate to customers how they could install water-saving devices, such as low-flow showerheads and water-efficient taps, in their homes. In addition the utility informed customers on the most water-efficient household appliances on the market. The result of the campaign was a reduction in water consumption in Berlin.

8.6.2   Today: Using water in the right way and reducing carbon emissions

Today BWB has stopped this programme because of rising groundwater levels due to lower consumption levels: when consumption decreased significantly (less groundwater was being withdrawn to meet supply), groundwater levels increased (the water table rose) so much that in many parts of Berlin today, houses/buildings cannot be constructed as the water table is too high – therefore, it is counterproductive for the utility to promote water conservation. Instead, BWB is focused on educating the public on how to use water in the right way, for example, educating customers on what they should not be putting into the sewage system.

When BWB's billing system sends an invoice out to customers once a year and advanced payment slips every 2–3 months, the utility often inserts a flyer providing customers with information. However, BWB has found that the more effective way of providing information to customers is through the Internet, and so the utility has a website where customers can, for example, check how much it will cost to have a new house connection. From time to time, BWB has image campaigns citywide for a device that adds CO_2 to tap water to create sparkling water. This is an attempt to save carbon emissions (and energy costs) associated with bottled mineral water: if Berlin drank tap water rather than mineral water, CO_2 emissions could be reduced by 99 000 tonnes per year – 1 litre tap water causes 0.35 gram of CO_2 emissions, while 1 litre of bottled mineral water causes 211 grams of CO_2 emissions. BWB carries out this campaign through the Internet and advertisements in the newspapers. BWB also provides an emergency service call centre that is responsible for all water-related issues including drinking water quality and pipeline bursts in Berlin. This call centre also has flyers in their cars to give to customers whenever they are called out. Finally, BWB surveys every 2 years customers on how satisfied they are with the utility, the tariff rates, etc. These surveys are important for BWB as it provides guidance on where the utility needs to perform better.

8.7   Case study SWOT analysis

8.7.1   Strengths

While BWB may not specifically promote water conservation, it still promotes the wise use of water by pricing it at its full economic cost. The price of water contains both a fixed and variable component ensuring revenue stability despite falling consumption: as water consumption decreases at 1.5 percent per annum, BWB should have sufficient revenue from the fixed component for the operation and maintenance of the water supply network.

While BWB has ceased promoting water conservation, the utility has instead focused on promoting the wise use of water along with the lowering of energy costs and carbon emissions. For instance, despite the utility's goal of reducing carbon emissions by 2010 being historical, BWB in 2012 installed wind turbines at its wastewater treatment plant to further reduce energy costs and reduce carbon emissions.

In line with using water wisely, BWB proactively invests significant amounts each year into its water system to reduce its UFW rate. This provides a good lesson to other cities that proactive, rather than reactive, management of UFW can lead to significant reductions in leakage.

BWB strictly enforces what types of activities can take place around its water wells to ensure source water is protected from contamination. This reduces the treatment and energy costs, as well as carbon emissions, of providing potable water.

8.7.2 Weaknesses

Due to increased environmental awareness, improved water technologies and renovations of apartments in Berlin, particularly in the eastern part of the city following reunification, reducing water consumption groundwater levels have risen significantly threatening the city's infrastructure from flooding.

While every customer in Berlin is metered, the actual population that is metered is very low due to the vast majority of Berliners living in rented apartments that do not require individual submeters. This inhibits the ability of the utility to communicate directly with all water users in Berlin, not only customers of BWB, on the wise use of water through communication tools such as billing inserts that raise awareness on issues, including what should not be flushed down the toilet or poured down the sink.

From an operational point of view, if consumption is too low, then BWB will have less revenue to cover the fixed costs of operating and maintaining the water distribution system. In addition, lower consumption levels result in not enough water to transport sewage to the treatment plants: BWB would have to use additional potable water to flush the sewers to maintain cleanliness of pipes, which is cost intensive. Even more cost intensive would be to reduce in diameter the sewage pipes.

Despite charging a fixed tariff for processing rainwater that flows off a customer's total sealed surface of their property, the utility does not provide economic incentives for customers inside the utility's combined system to minimise the total rainfall that needs processing at its wastewater treatment plants. Meanwhile on the outskirts of the city where there is a separated system, the utility does not provide incentives for customers to install rainwater harvesting systems or stormwater management systems to reduce the amount of potable water required or reduce contaminants entering waterways.

8.7.3   Opportunities

In the future, to promote the wise use of water and reduce the maintenance costs of the and wastewater system, the utility could either implement universal sub-metering of apartments or have BWB staff visit rented apartment complexes annually to distribute water-awareness materials. In line with BWB promoting the wise use of water, the utility is introducing AMRs to customers throughout Berlin; however the actual numbers of meters is low due to their high economic costs. Nonetheless, BWB could install AMRs with high users and publicise to the wider community how households have become more water wise following installation. In addition, the AMRs could also have software installed that calculates how much energy and carbon emissions are saved through using water wisely.

To reduce energy costs further, BWB should install more wind turbines and photovoltaic systems at its water treatment plants and explore ways of using renewable energy at its well sites. In turn this will show consumers that BWB is a responsible corporate citizen 'doing its part' to protect the environment by using resources wisely.

Because of the need to protect groundwater supplies located inside the city from contamination, rainwater harvesting systems that could be used to reduce the economic costs of treating wastewater are discouraged. Within the combined system, BWB could encourage customers to develop, through rebates, stormwater urban design systems (SUDS) that capture rainwater from sealed surfaces for filtration through natural vegetation before replenishing groundwater levels. In specific areas where groundwater recharge is not possible, BWB could encourage customers to develop SUDS retention ponds that store rainwater during extreme storm events before allowing it to be released into the combined system for treatment, therefore reducing the volume of wastewater needing to be treated during wet, storm events, which in the long term reduces operational and infrastructural costs to BWB from building larger-capacity wastewater treatment plants. Alternatively, BWB could promote rainwater harvesting devices that are made of approved materials and work with distributors to encourage only approved ones to be sold.

8.7.4   Threats

Berlin will face challenges to its water resources in the future from climate change with increased precipitation levels in winter and decreased levels in summer. Heavy rainfall can lower the availability of good quality water due to contamination of groundwater supplies during storm events. Meanwhile, the utility will need to ensure that the wise use of water becomes culturally ingrained as there is the potential for droughts over warmer periods. Berlin may have to consider the use of greywater and rainwater harvesting as sources of alternative supplies during warm, drier months. In addition, with falling water consumption levels, water and wastewater may become stagnant leading to potable water being flushed through the system. If this were the case, BWB would have to consider the use of smaller diameter pipelines in the water system; however the capital cost in replacing the present pipelines is enormous.

8.8   Transitioning towards urban water security summary

BWB uses a portfolio of demand management tools to achieve urban water security (Table 8.2). However, there are numerous barriers identified by the utility in achieving further urban water security in Berlin (Table 8.3).

Table 8.2 Demand management tools to achieve urban water security

Diffusion mechanisms	Tools	Description
Manipulation of utility calculations	Pricing of drinking water and wastewater	Water priced at its full economic cost
		Customers are charged for processing sewage and rainwater
Legal and physical coercion	Metering	Universal metering for both domestic and non-domestic customers
		No submetering of rented apartments in property company-owned buildings
		BWB is introducing AMRs in the city
Socialisation	Water-efficient technologies	Public demonstrations of how water-saving devices can be installed in homes
		Tap filters that add gas to tap water turning it into sparkling water
Persuasion	Public education	Kindergarten and school visits
		The Internet providing information on water issues
		Water-saving campaigns in newspapers
		Travelling exhibitions
		Information on water-efficient appliances
		Billing inserts to inform the public on how to use the water system in the right way
		Brochures in BWB vehicles
		BWB's call centre provides advice to customers
		Promotion of tap filters that turn tap water into sparkling water

Table 8.3 Barriers to further urban water security

Barrier	Description
Economic	If consumption decreases further BWB will have less revenue to cover the costs of operating and maintaining the water distribution system
	AMRs are not distributed widely due to their high economic costs
Infrastructural	If consumption is too low there is not enough water to transport sewage to the treatment plants, so additional potable water is required to flush the sewers to maintain cleanliness of pipes
Regulatory	BWB does not utilise greywater because of the potential for cross-connections
	BWB does not promote rainwater harvesting because many of the rainwater traps on the market in Germany are made out of copper, increasing copper levels in the groundwater, which could lead to all rainwater in the future requiring treatment (economic costs)
	Despite BWB providing water to around 3.5 million inhabitants of Berlin, the utility only has around 250 000 customers due to no regulations on submetering of rented apartment complexes

Notes

1. SWTICH – MANAGING WATER FOR THE CITY OF THE FUTURE. 2010. Making urban water management more sustainable: achievements in Berlin [Online]. Available: http://www.switchurbanwater.eu/outputs/pdfs/w6-1_gen_dem_d6.1.6_case_study_-_berlin.pdf (accessed 11 May 2016).
2. Ibid.
3. Ibid.
4. BERLIN CLIMATE PROTECTION INFORMATION OFFICE. 2015 [Online]. Available: http://www.berlin-klimaschutz.de/en/berlin-water-act-bwg (accessed 11 May 2016).
5. BERLINER WASSERBETRIEBE. 2013. Water for Berlin: clear water clear information. Berlin, Germany: Berliner Wasserbetriebe.

9

Copenhagen transitioning towards urban water security

Introduction

Copenhagen's transition towards urban water security focuses on reducing leakage, protecting the city's groundwater sources and changing behaviour to reduce wasteful consumption. This case study analyses how Copenhagen's water utility uses a portfolio of demand management tools to modify the attitudes and behaviour of water users to achieve urban water security.

The case study first provides a brief company background of Copenhagen's water utility, along with an overview of the city's water supply and water consumption levels before discussing the city's strategic vision for achieving urban water security. The case study will then analyse the various demand management tools used by the city's water utility in an attempt to achieve urban water security before discussing the numerous barriers identified by the utility in achieving further urban water security in Copenhagen.

9.1 Brief company background

The water supply in Copenhagen was managed by the city until 2005, when it was separated from the city and became a private company – Copenhagen Energy – with all the stocks owned by the city of Copenhagen. In 2012, Copenhagen Energy merged with seven other water suppliers around Copenhagen to form HOFOR. HOFOR is the largest utility in Denmark providing 20 percent of the Danish population with water supply, wastewater management and other services. Its turnover is

Urban Water Security, First Edition. Robert C. Brears.
© 2017 John Wiley & Sons, Ltd. Published 2017 by John Wiley & Sons, Ltd.

around EUR 500-600 million a year and the utility employs around 800 people. The city of Copenhagen owns 73 percent of the company with other municipalities around Copenhagen owning the rest.

9.2 Water supply and water consumption

HOFOR provides drinking water to approximately 550 000 consumers in Copenhagen from a water network consisting of 15 waterworks, 57 well fields and 540 abstraction wells. Only 4 percent of Copenhagen's groundwater supplies come from within the city's administrative/political boundaries, the rest are 'imported' from other municipalities.

HOFOR provides drinking water sourced from groundwater abstracted in the large area of Zealand. The utility abstracts around 50 million cubic metres of water annually with 49 well fields supplying water to seven regional water companies and eight well fields supplying local waterworks. The vast amount of groundwater abstracted can be used for drinking without treatment other than oxidation and filtration. However, one waterworks (Hvidovre) has an additional water treatment process with a carbon filter ensuring there are no residues from herbicides in the drinking water supplied, while another treatment plan (Store Magleby Waterworks) has a UV plant disinfecting the water.

Currently, water consumption in Copenhagen is 104 litres per person per day. Sixty-nine percent of HOFOR's customers are domestic, while 27 percent are non-domestic. Household water consumption in Copenhagen has steadily declined over the past almost 30 years from 174 litres per person per day in 1985 to 104 litres per person per day (Table 9.1).

9.3 Strategic vision: Water supply plan (2012–2016)

In Copenhagen, the city council's water supply plan is developed by the Centre for Environment, a department of the Technical and Environment Administration, in collaboration with HOFOR. The water supply plan contains the limits and

Table 9.1 Distribution of water consumption types

User	Percentage
Household	67
Profession	21
Institution	6
Unmetered	4
Leisure	2

conditions together with lessons learnt from past efforts and constitutes the basis for the establishment of the city's specific water conservation targets with the setting of the targets based on security of supply, good water quality and reduction in water consumption of imported water. In addition, the water plan also details which groups of water users the city would like to see lower their water consumption. The budget HOFOR receives from the city to implement this water plan is around 1.9 million DKK. The funding is not conditional on implementing specific water conservation campaigns; instead HOFOR has the freedom to target different water users in different years to achieve the overall water consumption goals. However, the utility is not expected to achieve specific results but instead has to show they are proactive in promoting water conservation. The city of Copenhagen continuously revises the water supply plan every 4 years ensuring initiatives are taken for reducing water consumption across the city.

The new water supply plan (2012–2016) aims to achieve a domestic water consumption target of 100 litres per capita per day and 301 litres per capita per day for non-domestic users by 2017 and a 13 percent reduction in water used per day per citizen from 100 litres in 2017 to 90 litres per day in 2025.[1] In addition, HOFOR has an alternative water supply target that 4 percent of water consumption should be from secondary water sources by 2017. Regarding specific water users the city wishes to see more water savings in children's institutions, for example, kindergartens and schools for water conservation efforts. To achieve the water reduction targets, HOFOR's strategy is to monitor and prevent leaks, protect groundwater sources and change behaviour through water meters and pricing mechanisms to reduce wasteful consumption.[2]

9.4 Drivers of water security

The drivers of HOFOR's strategic vision for achieving urban water security include variations in the availability of good quality water of sufficient quantity as well as the political dimensions of importing water (summarised in Figure 9.1).

9.4.1 1980s: Quantity of water

In the 1980s, Copenhagen had to resort to abstracting water from lakes as demand for water outstripped supply. This led to the chlorination of drinking water as the lake water quality was not of spring-like quality compared to groundwater

Figure 9.1 Evolving drivers of water security in Copenhagen.

supplies. As such, the taste shifted from fresh spring water to industrially treated water. This resulted in consumers being unhappy with the taste of the water. In addition, a segment of society was concerned with the environmental impact of withdrawing water from the lakes.

9.4.2 1990s: Quality of water

In the 1990s the range of chemicals that needed to be tested for in drinking was expanded. The result of the new drinking water quality standards was the closure of numerous wells the city depended on for supply – not because of the water being dangerously contaminated but due to the inability of the utility to treat water with chlorine. This led to an actual quantity problem in 1993 and 1994 when Copenhagen was close to not having enough wells to meet the demand. At the same time the government implemented a tax on water to lower consumption.

9.4.3 2000s: Political and quality of water

Between 2000 and 2010, Copenhagen's permits to transport water from neighbouring cities expired. However, the cities that were providing water to Copenhagen were reluctant to renew these permits as they wished to retain the water for their own consumption. In addition, the quality issue reappeared with the updating of water quality testing standards, which increased in scope the number of chemicals needing to be tested for. Therefore, the city faced the possibility of further wells being closed down.

9.4.4 2010 onwards: Quality and quantity of water

From 2010 onwards water conservation has been driven by quality issues with additions to the number of chemicals being tested for in groundwater supplies including round-up. This could potentially lead to further closing of wells Copenhagen relies on for groundwater supplies. In addition, climate change in the long term will pose a threat to Copenhagen's water supplies with spatial and temporal changes to precipitation levels. Specifically, precipitation is projected to increase by 25–55 percent during the winter and decrease by up to 40 percent in the summer while heavy downpours, which typically occur in late summer, will become 30–40 percent heavier. Increased storms could contaminate groundwater supplies from surface runoff, while longer periods of drought between heavy precipitation events could reduce groundwater recharge. In addition, more frequent and more intense heatwaves in the future could lead to increased demand for water. In addition, demand for water will increase with rapid population growth in Copenhagen: the city's population is projected to grow by an additional 100 000 people, from 535 000 in 2010 to 640 000 in 2025 – a result of a 30 percent increase in young people (20–29 years) and a 50 percent increase in elderly (65+ years) over the period 1980–2024. As a result, the number of housing developments will increase with 45 000 new apartments to be built over the next 25 years.[3] In the long term, HOFOR expects to have water scarcity issues. Agricultural production has polluted groundwater and climate change will bring long-term dry spells; therefore, HOFOR is determined that all citizens have to save water and use it efficiently. Groundwater supplies are threatened by pollution mainly from residues from pesticides and chlorinated solvents. Around 10 percent of HOFOR's wells have closed due to groundwater pollution.

9.5 Regulatory and technological demand management tools to achieve urban water security

9.5.1 Pricing of water and wastewater

The price of water HOFOR charges is strictly regulated by the local government: any price change requires local government approval based on the 'non-profit principle'. This ensures the utility cannot use the price of water for revenue generation as citizens have a right to water, and therefore it cannot be priced too high. As such, the price of water in Copenhagen only covers the full economic costs of providing the water. In 2010, Denmark's competition and consumer authority determined that the price of water in Denmark is too expensive. This effectively placed a price ceiling on water with all Danish water supplies. At the same time, the government passed a law stating that all Danish water utilities need to become more efficient; specifically each utility must reduce, by 2018, their operating budgets by 25 percent. The pricing structure of water in Copenhagen includes both fixed and variable components (Table 9.2). In 2013, the total price of water was 39.11 DKK per cubic metre of water (1000 litres).

HOFOR charges domestic and non-domestic customers a volumetric rate of 22.80 DKK per cubic metre of wastewater. In addition, there are two fixed tariffs for covering the costs of transporting and treating wastewater. Table 9.3 provides

Table 9.2 Water tariffs for domestic and non-domestic customers

Price composition per cubic metre of water	DKK
Water tariff (DKK per cubic metre)	6.42
Groundwater protection	0.50
Water tax	5.46
State tax for mapping groundwater resources	0.67
Drainage contributions, transport	6.38
Drainage contributions, cleaning	11.86
VAT	7.82
Total drinking water cost	**39.11**

Table 9.3 Wastewater tariffs for domestic and non-domestic customers

Price composition per cubic metre of wastewater	DKK
Drainage contribution to transport	6.38
Drainage contribution to cleaning	11.86
VAT portion of wastewater	4.56
Total wastewater cost	**22.80**

Table 9.4 Additional wastewater tariffs for non-domestic with high pollution content

Additional charges	DDK (excl. VAT)	DKK (incl. VAT)
Suspended solids (per kilogram)	1.40	1.75
Nitrogen (per kilogram)	9.24	11.55
Phosphorus (per kilogram)	39.16	48.95
Administrative fee (per company)	20 000.00	25 000.00

a breakdown of the wastewater charge. Non-domestic customers with higher pollution content than ordinary domestic wastewater pay additional costs for the removal of suspended solids, nitrogen and phosphorus. In addition, these customers pay an administrative fee relating to public sewage plant construction and operation (Table 9.4).

9.5.2 Metering

Under Danish legislation all properties – domestic and non-domestic – connected to common waterworks must have water meters installed at the property level. Properties with several flats are however only required to install one water meter at the property level. In Copenhagen, the city council has agreed to allocate HOFOR DKK 2 million annually to support the installation of individual water meters, in addition to water-saving toilets, in housing association buildings, as about 90 percent of housing association buildings do not have individual water meters: the utility's research has found that housing association buildings make water savings of around 20 percent after installing individual meters.[4]

9.5.3 Reducing unaccounted-for water

Copenhagen's water supply plan (2012–2016) aims to ensure UFW is less than 10 percent through continuous renewal of the water distribution system,[5] a challenge given the city experiences harsh winters and some pipes date back to the

1860s and 1870s. With frosts over winter the utility naturally sees a spike in leaks during the spring. Currently, HOFOR's UFW rate is around 7 percent due to an active leak detection and network rehabilitation programme.

Active leak detection

HOFOR has around 1100 kilometres of pipes, and the utility checks for leaks throughout the whole system over a rolling 4-year period. Since 2012, the distribution grid is systematically checked over a rolling 3-year period. HOFOR also uses loggers that are placed between fire hydrants at night to listen to leaks.

Network rehabilitation

Since 1929, Copenhagen's pipe failures have been registered with information about dates, location and reason for failure. HOFOR has used registration of failures to prioritise the areas and types of pipes that need renovation to prevent future failures. HOFOR aims to renovate the distribution network at a renewal rate of 1 percent annually, corresponding to the utility renovating 9 kilometres of water pipes annually. However, there are great differences in the price of renovation of the different parts of the distribution grid. In the last water supply plan (2008–2012), renovation of around DKK 34 million was carried out over the 4-year period, resulting in 6.8 kilometres of water pipes being renovated. In the future, the goal is to renovate the distribution grid so UFW is kept as much under 10 percent as technically and financially possible.[6]

9.5.4   Source protection: New forests and reducing pesticide use

HOFOR is in the process of planting new forests together with local municipalities, other water utilities and the Nature Agency to protect groundwater sources. The planting of the new forests began in 2002, and the total size of the forested areas when completely planted will be approximately 4000 acres, equivalent to 8000 football fields. Afforestation ensures there are no pesticides or deposited sludge, soil or other waste products contaminating groundwater. Naturstyrelsen buys land and plants trees and owns the forest and is responsible for its management. In addition to different deciduous trees, there are also green meadows, grasslands and lakes that help preserve nature and ensure rich animal and bird life.

HOFOR has established voluntary cultivation agreements with farmers and planning of forests in vast catchment areas. In 2010, the planted area was 557 hectares and the total project for afforestation is 3840 hectares. Part of the voluntary cultivation agreement is that the farmers do not use pesticides on their property. The farmer receives compensation for the income input when cultivating areas without pesticides. If the farmer has land where the municipality has designated an afforestation area, the utility can provide compensation for the planting of trees without the use of pesticides. Meanwhile, inside the city's boundaries, HOFOR participates in the Vestegnens VandSamarbejde water cooperation project, which involves encouraging gardeners to stop using pesticides.

9.5.5 Developing alternative water supplies

The city has a target of 4 percent secondary water. Nonetheless, the frame of interpretation for secondary water is very wide while the permits are narrow. Greywater (reuse of water) is strictly prohibited in Copenhagen. In the 1990s, there was a housing project that had a greywater project attached to it through the issuance of a temporary permit, which has expired 5–7 years ago. The municipality will not renew the permit, but the plant is still operational so they will not close it down, that is, the municipality has turned a 'blind eye' to it. HOFOR receives numerous requests from interested companies along with private housing companies asking for funding towards greywater projects; however despite HOFOR being willing to co-fund the projects, they will not be approved by the municipality.

Very few large users of water request permits to use water of drinking water quality for industrial purposes because they are usually rejected for the reason that drinking water is only for consumption. If a permit is approved, it is only for the purpose of using water as a coolant, for example, a bank made a request to use drinking water for the cooling of their computer servers, but this application was initially rejected; however, as the bank was only going to use the water for emergency purposes, as a backup, a comprise was made in that if any water is used the bank would have to retain that water for reuse at a later date, reducing the amount of water being drawn from the water supply in any future emergency.

Frequently interested companies, including private housing companies, contact HOFOR about installing greywater systems and whether they can apply for funding to do so. Despite the utility being willing to facilitate the installation of greywater systems, including co-funding, the projects will not receive the required permits from the local authorities. A local company producing industrial gases required water for cooling purposes and had developed a private wells; however, city authorities have to approve all private wells, and they will only be approved if the water is of drinking water quality – and drinking water is only for potable use.

Meanwhile, it is only over the past 10 years that rainwater harvesting has been allowed in Denmark; however, the systems can only be installed for flushing private toilets. To install this system, a permit from the city is required for installation. In addition, HOFOR has to control the backflow valve twice a year (for free). For housing association buildings, the regulatory paperwork is more burdensome because, unlike private houses where only a few members may become ill from contaminated water, these buildings have hundreds of tenants. HOFOR was involved in a project related to the use of rainwater in housing complexes involving 30–40 buildings. After 4 years, one plant was a private well with water of drinking quality, and the remaining two were a combination of water from a well and rainwater collected from the roof. However, the burden of 'red tape' meant that over the 4-year period the special dispensation to conduct the experiment had to be periodically renewed, and eventually the project was ended due to this regulatory burden. The price for rainwater harvesting systems (for flushing private toilets) is around 50 000 DKK (constructed of approved materials and approved by the authorities) with a payback period of 30–40 years. The result is the average consumer will not have one installed.

HOFOR believes that the aversion to alternative sources of water is due to the cultural perception that groundwater is 'holy' as it is of spring quality and does not require treatment before drinking. In addition to potentially contaminating groundwater from alternative sources, the authorities believe that if society started to accept water of poor quality, which requires treating, then there would be no incentive to protect the groundwater in the first place. In addition, the public demands untreated, spring-quality water: when HOFOR was forced to use UV treatment at one of its storage facilities, the public deemed this to be a failure as the water was of poor quality despite it being a problem with the facility itself.

9.5.6   Reducing energy costs and carbon emissions

In order to limit carbon emissions from the operation of pumps, HOFOR has set up wind turbines and photovoltaics in a single-well field. From 2013, HOFOR will erect a further 100 wind turbines at other source sites.

9.5.7   Subsidies for toilets and water meters

HOFOR provides housing associations with a toilet subsidy. Under this scheme, housing association buildings are eligible for 1000 DKK per apartment to replace toilets with more efficient ones with the subsidy being a quarter of the price of a new water-efficient toilet. Housing association buildings are also eligible for a 1000 DKK subsidy for the installation of individual water meters for each apartment. The total budget for toilet and water meter subsidies is 2 million DKK per year. HOFOR also has a subsidy scheme for rainwater harvesting systems in private homes and industrial projects with a total budget of 750 000 DKK.

9.5.8   Consultants and water conservation advice

Commercial water users can request for free a consultant from HOFOR for advice on water conservation. However, these requests are infrequent as commercial customers are extremely competent in identifying and solving issues as they have their own technical staff. However, housing associations and private consumers often ask for advice. If private homes require assistance in reducing their water consumption, HOFOR will send out an employee to visit the customer and help look for ways to save water. However, frequently, water consumption in private homes is high because the homeowner either does not care and has a high income to not worry about the water bill or does not know there is a problem and therefore does not contact the utility to help reduce water consumption levels. If a customer has a sudden increase in water consumption – a 10 percent change in water consumption – the next water bill will automatically contain a letter informing the customer of this increase and to look for leaks.

9.5.9   Water-saving devices

HOFOR often distributes flow reducers for taps for free to whoever requests them particularly housing associations, as they are aware their consumption is high and wish to lower it. If housing associations contact HOFOR, the utility often sends out around 500 flow reducers for installation by housing association employees. HOFOR keeps abreast of water-efficient technology updates and shares them with customers. At times, the utility may even hand out samples; however, this is limited by the water conservation budget. Frequently, local inventors approach HOFOR to test their water-saving devices. The utility then finds test families and has the devices installed in their homes to collect data before writing up a report containing the results. Customers also frequently ask HOFOR to recommend the best water-saving devices on the market. However, as the utility is a monopoly, it is careful not to endorse specific products; instead, it directs customers towards consumer product tests and recommends the customer compare the top three products.

9.6   Communication and information demand management tools to achieve urban water security

9.6.1   Education and awareness in schools

HOFOR runs a school programme called 'water heroes' in which a utility employee visits grade 1 or grade 2 classes for around three hours teaching children about water conservation. During this time the class nominates one male and one female water hero and a villain (male or female). Normally the utility visits less than 10 classes per year, but in 2012 the utility enabled more to enrol in the programme and 34 classes signed up. Through this programme the utility also targets the parents as each child receives a water book to take home along with their drawings of their water heroes. In addition, a website teaching programme has been prepared comprising teaching of water and drainage for pupils in basic school and upper secondary schools.

9.6.2   Public education

HOFOR has a Facebook page in which the utility at times runs competitions with the winners receiving water-efficient devices such as showerheads. HOFOR will hire a university researcher to conduct a study on how the utility can target different demographic groups effectively. Technically, the utility has expertise in creating water conservation campaigns, but to date these have been too broad, and now the utility aims to target specific water user groups with tailored messages. Regarding protecting groundwater resources, Vestegnens VandSamarbejde has conducted groundwater protection campaigns involving distribution of information, television advertisements, social media campaigns and articles in magazines to encourage people not to use pesticides. The media also play a role in promoting environmentalism with the media frequently running stories on water conservation even if HOFOR has not provided a story. Stories are often about families that have successfully saved water or a company that decided to implement water-efficient practices and saved a considerable amount of water.

9.6.3 Challenges of public awareness campaigns

HOFOR cannot charge the price they would like to in order to promote further water conservation, as there is a price ceiling on how much the utility can charge for water. As such, water conservation campaigns are limited by a small campaign budget, and so the utility cannot initiate large outdoor water conservation campaigns and TV commercials. HOFOR does not use billing inserts to promote water conservation due to political resistance as well as the cost of writing letters to all its customers. It is extremely difficult to explain to people that despite the frequency of rain, particularly in the summer of 2012 when there was heavy precipitation levels, people still need to conserve water despite the landscape being lush and green. The reasoning is that falling precipitation is mostly recharging groundwater which is polluted from agricultural runoff along with residue from industrial pollution; therefore, the availability of good quality groundwater of sufficient quantity is limited. This issue will only become worse over time; however, HOFOR has difficulties in framing the message in a way that people understand yet ensure it does not sound alarmist. It is always the 'green' people who seek advice from HOFOR. The real target is the people who do not know they have a problem, and they are usually the people in housing association flats that do not have water meters.

9.7 Case study SWOT analysis

9.7.1 Strengths

Regarding the water plan, HOFOR decides on which sectors it will specifically target with its limited budget and the methods to be employed. This means a concentrated effort can be made on achieving water savings within specific sectors rather than stretching a small budget thinly over many sectors and achieving limited results.

On the issue of exposure to falling revenues from lower consumption levels, HOFOR's water tariff includes both fixed and variable components: the fixed components provide some revenue stability towards operating and maintaining the water supply network effectively.

To cover energy and other related costs of treating wastewater, HOFOR charges non-domestic customers with higher pollution content additional costs for the removal of suspended solids, nitrogen and phosphorus.

HOFOR recognises that housing association buildings are large consumers of water. With toilets being one of the largest uses of water in a household, the utility has initiated a toilet retrofit programme where old toilets are replaced with new ones with HOFOR subsidising a quarter of the price.

Copenhagen has a very low UFW rate of just 7 percent despite harsh winters and a reliance on old concrete water pipes, some of which are 130 years old. This is because the utility has initiated an active leak detection programme that surveys every pipeline throughout the city over a 3-year period.

9.7.2   Weaknesses

HOFOR faces a price ceiling on how much it charges consumers for water, and so it must rely on non-price demand management tools to promote water conservation to achieve its water consumption target.

HOFOR faces regulatory hurdles on promoting rainwater and greywater harvesting systems to meet the goal of 4 percent of water consumed coming from secondary water resources.

Promoting water conservation in housing association buildings is made difficult by the fact that buildings do not have submeters in each apartment which the utility recognises as being a challenge given that water meters encourage significant reductions in water consumption levels.

Non-domestic customers cannot drill private wells for the purpose of using raw water in industrial/commercial operations because the city will only approve wells if the water is of drinking water quality.

9.7.3   Opportunities

While HOFOR subsidises the installation of water meters in housing association buildings, the utility needs to prioritise this programme to ensure water conservation targets are achieved. For instance, the budget for rainwater harvesting could be diverted towards the installation of water meters to achieve universal metering in the future. In addition, submetering would enable the utility to directly communicate with water users in each apartment rather than through the building owner.

With more investment HOFOR could reduce its UFW rate further with higher customer feedback on leaks in addition to the utility conducting more frequent surveys of the water system.

The utility could combine its current service of offering visits to private homes to find ways of conserving water with its automated system that informs customers of a sudden increase in water consumption. Specifically, when an automated letter is sent out, the utility could offer the customer the option of a free visit by a HOFOR representative. In addition to providing water conservation tips, the representative could provide information on the latest water-efficient devices on the market. HOFOR could also send employees to housing association buildings to provide water-saving tips to occupants and professionally install tap inserts.

HOFOR will first need to conduct demographically targeted water conservation programmes in the future which means water conservation messages are tailored to specific demographic groups increasing the likelihood of water savings, and, second, the utility could identify more accurately specific high water-consuming groups for the distribution of water-saving devices.

HOFOR could utilise not-for-profits to address specific demographic groups outside of the water plan's targeted sectors, enabling the utility to achieve increased water savings across the board.

HOFOR can utilise the local media to publicise how individuals and households have successfully saved water. For instance, HOFOR could publicise winners of social media competitions along with stories on how they came up with winning water-saving ideas, while water heroes, nominated by school classes, could be advertised as models for other young people to emulate.

As HOFOR already offers non-domestic customers a consultancy service, the utility could develop a water audit programme for its customers, enabling the utility to conduct case studies on how non-domestic customers have saved water and have the media publish these stories for other businesses to emulate.

HOFOR has difficulties promoting water conservation during periods of heavy rainfall when vegetation is lush. HOFOR could include in its school water heroes programme competitions for the best posters that inform people on the need for conserving water year-round. When HOFOR conducts retrofits of toilets, it could also provide housing association residents with information brochures on year-round water conservation.

9.7.4   Threats

Copenhagen lacks good quality water of sufficient quantity from agricultural and industrial pollution of groundwater supplies. With increased standards of chemical testing, Copenhagen faces the possibility in the future of a decrease in supply from wells being closed down.

Politically, nearly all of Copenhagen's water is imported from outside the city. With neighbouring cities wishing to use the water for their own consumption, Copenhagen must be seen as a responsible consumer of water to ensure water permits are renewed in the future.

Economically, the government of Denmark has required all Danish water utilities to decrease, by 2018, their operating budgets by a quarter, resulting in the need to reduce operational costs of providing water services. This will place fiscal stress on water conservation budgets, making it difficult for the utility to achieve its water consumption targets.

Climate change is expected to lead to increased groundwater levels and changing precipitation patterns. This may lead to contamination of drinking water due to leaking drinking water pipes.

9.8 Transitioning towards urban water security summary

HOFOR uses a portfolio of demand management tools to achieve urban water security (Table 9.5). However, there are numerous barriers identified by the utility in achieving further urban water security in Coponhagen (Table 9.6).

Table 9.5 Demand management tools to achieve urban water security

Diffusion mechanisms	Tools	Description
Manipulation of utility calculations	Pricing of drinking and wastewater	Water is priced to recover the full economic costs of providing water services
		Customers are charged for processing wastewater
		Heavy non-domestic users are charged for removing waste products and a fee for maintaining the system
	Subsidies for toilets, water meters, and rainwater systems	Subsidies for housing association buildings to install water-efficient toilets and water meters
		Subsidies for the installation of rainwater harvesting systems for flushing private toilets
Legal and physical coercion	Metering	Every domestic and non-domestic customer has an individual water meter. However, 90 percent of housing association buildings do not have submeters
	UFW	There is an active leak detection programme with the network systematically checked for leaks
	Developing alternative sources of water	Majority of secondary sources to date is from rainwater harvesting systems for flushing private toilets
Socialisation	Water-saving devices	Distribution of water-saving devices
	Private household visits and information	Private household visits to help homeowners find ways of reducing water consumption
		Customers are informed of sudden increases in water consumption, suggesting a possible leak
Persuasion	School education	Water heroes in classrooms
	Information on water-efficient devices	Customers are provided information on new water-efficient products (devices and appliances) available on the market
Competition, lesson drawing and emulation	Competitions	Facebook competitions for water-saving devices
	Classroom water heroes	The 'water heroes' school programme involves classrooms nominating a male and female water hero and a water villain
		Children then take home their classroom materials and drawings of their water heroes to show to their parents
	Reducing energy costs	Wind turbines and photovoltaics have been installed in a single-well field
		Planned installation of 100 wind turbines at other source sites in the future

Table 9.6 Barriers to further urban water security

Barrier	Description
Regulatory	HOFOR cannot increase the price of water to encourage water conservation as there is a price ceiling on how much customers can be charged
Economic	A small conservation budget limits the types of outdoor water conservation campaigns and TV commercials
Regulatory	Rainwater harvesting systems require a lengthy application
	Greywater is strictly prohibited in Copenhagen due to the potential of cross-contamination
	Non-domestic customers cannot drill private wells for the purpose of using raw water as wells should be of drinking water quality
Framing	It is extremely difficult to explain to people that despite periods of heavy rain they still need to conserve water
Cultural	Copenhagen and Denmark are very proud of their clean groundwater so authorities are reluctant to approve experiments involving reusing water due to the potential for groundwater contamination
Demographic	It is always the 'green' people who seek advice from HOFOR

Notes

1. EUROPEAN COMMISSION. 2014. European green capital award – Copenhagen [Online]. Available: http://ec.europa.eu/environment/europeangreencapital/winning-cities/2014-copenhagen/ (accessed 11 May 2016).
2. CITY OF COPENHAGEN. 2014. Copenhagen: solutions for sustainable cities [Online]. Available: http://kk.sites.itera.dk/apps/kk_pub2/pdf/1353_58936BnEKE.pdf (accessed 11 May 2016).
3. CITY OF COPENHAGEN. 2010. Urban planning in Copenhagen: towards a sustainable future [Online]. Available: http://siteresources.worldbank.org/ECAEXT/Resources/258598-1279117170185/7247167-1279119399516/7247361-1279119430793/urbandev.pdf (accessed 11 May 2016).
4. EUROPEAN COMMISSION. 2014. European green capital award – Copenhagen [Online]. Available: http://ec.europa.eu/environment/europeangreencapital/winning-cities/2014-copenhagen/ (accessed 11 May 2016).
5. Ibid.
6. Ibid.

10 Denver transitioning towards urban water security

Introduction

Denver's transition towards urban water security focuses on lowering total water consumed by its customers. This case study analyses how Denver's water utility uses a portfolio of demand management tools to modify the attitudes and behaviour of water users to achieve urban water security.

The case study first provides a brief company background of Denver's water utility, along with an overview of the city's water supply and water consumption levels before discussing the city's strategic vision for achieving urban water security. The case study will then analyse the various demand management tools used by the city's water utility in an attempt to achieve urban water security before discussing the numerous barriers identified by the utility in achieving further urban water security in Denver.

10.1   Brief company background

Denver Water serves more than 1.3 million people in Denver and its surrounding suburbs. While it is an agency in the city, the utility is independently run with its own governing board. The utility only supplies water while other city agencies manage wastewater and stormwater. Denver Water serves around a quarter of the state's population but uses less than 2 percent of all water, treated and untreated, in Colorado. Its mission statement is that the utility will be a

Urban Water Security, First Edition. Robert C. Brears.
© 2017 John Wiley & Sons, Ltd. Published 2017 by John Wiley & Sons, Ltd.

responsible steward of the resources, assets and natural environments in order to provide a high-quality water supply, a resilient and reliable system and an excellent customer service.

10.2 Water supply and water consumption

In the State of Colorado, water that flows out of the state travels to either the Atlantic or Pacific oceans, depending on which side of the Continental Divide the water originates: on average, 10 434 000 acre-feet of water leaves the state each year (1 acre-foot is equal to 325 851 gallons of water and will supply slightly more than two single-family households for a year). About 80 percent of the water in Colorado is found on the West Slope, but about 80 percent of the state's population lives on the East Slope that includes Denver.[1] That division means the transferring of water from the West Slope to the East Slope through trans-basin diversions. Utilities across the East Slope transfer about 475 000 acre-feet of water from the Colorado River basin to the East Slope each year. On average, Denver Water's customers use about 125 000 acre-feet of West Slope water per year. Interstate compacts regulate how much water needs to flow from Colorado to downstream states. Two interstate compacts directly affect river systems from which Denver Water derives its supply: the Colorado River Compacts of 1922 and 1948 and the South Platte River Compact of 1923. The Colorado River Compact of 1922 divided the Colorado River into upper and lower sections. The dividing point is Lee's Ferry, which is located near the Utah and Arizona state lines. The compact requires that the upper basin states deliver 75 million acre-feet of water to the lower basin over any 10-year period. The South Platte River Compact of 1925 determines the amount of water that must flow from Colorado to Nebraska.[2]

Denver Water uses gravity to provide water to approximately 60 percent of its potable water customers. The remaining 40 percent rely on pump stations to deliver them water. There are two types of pump stations. The first type lifts water from lower elevations to fill treated water reservoirs at various high points around Denver. From there either gravity takes over to supply customers downhill, or it may be pumped uphill to yet another reservoir. For other areas where reservoirs are not an option, booster pump stations ensure adequate pressures are maintained at all times.

Denver Water has 18 potable, three recycled and two raw water pump stations in various locations throughout the distribution system, with a capability of pumping more than 1 billion gallons (Table 10.1). The utility does not use all of the 124 pumps in the system at the same time; rather they cycle them on and off based on the need. Some pump stations will always have pumps running, while others may only be needed in the height of the summer water-use season. Denver Water also has pumps that automatically increase and decrease with shifts in water demand, and many stations in the system can pump to two different elevations.

Table 10.1 Denver Water's infrastructure

Water infrastructure	Statistics
Miles of water mains (pipelines)	Over 3000
Miles of non-potable pipes in the system	45
Number of pumping stations	18 potable, three recycled and two raw water
Underground reservoirs in various city locations	30

Table 10.2 Total retail treated water use by category

Category	Percentage
Single-family homes	48
Business and industry	23
Multifamily homes	19
Irrigation only	6
Public agencies	4

10.2.1 Recycled water

Denver Water's source water for the recycling plant is treated wastewater from the Robert W. Hite Wastewater Treatment Plant. The recycling plant's current capacity is 30 million gallons per day (MDG) and is expandable to 45 MDG. The distribution system includes more than 70 miles of pipe with two major pumping stations and dedicated storage facilities. Denver Water provides recycled water to more than 80 customers, including parks and golf courses, the Denver Zoo, schools, homeowner associations and industrial complexes. Each year the utility provides 7000 acre-feet of recycled water for irrigation and industrial and commercial operations that do not require drinking water. Over the next 10–15 years, the recycled water service area will be expanded – to Denver International Airport and through central Denver – with Denver Water aiming to deliver 17 500 acre-feet of recycled water each year, freeing up enough drinking water to serve more than 43 000 households.

10.2.2 Customer segments

Denver Water's residential customers use 82 gallons a day per person for all indoor and outdoor purposes, down from 104 gallons in 2001. Nearly 70 percent of its customers are residential with the remaining 30 percent split between commercial, industrial and government, which include the city's large park system (Table 10.2).

10.3   Strategic vision: Denver Water's 22 percent water target

In 2006, the Denver Water Board of Commissioners set a water conservation goal of reducing water use to 165 gallons per person per day by 2016 – a 22 percent reduction from average pre-2002 drought use of 211 gallons per person per day – where gallons per person per day is calculated using total treated water consumed by Denver Water's customers divided by the population in the service area and includes all users: residential, commercial, industrial and institutional. Since the introduction of the 22 percent water target, Denver Water has declared an additional target: 30 gallons a day for indoor use.

10.3.1   Denver Water environmental stewards

The utility serves the whole city as well as customers outside the city under contract. The utility has various arrangements with its customers outside of the city – sometimes Denver Water provides the whole service or is just a wholesaler – there are 65 providers outside the city. If they receive water from Denver Water under one of these distribution arrangements, they have to run a conservation plan because Denver Water's 22 percent reduction goal relies on all its customers reducing water consumption levels. Denver Water has made water conservation part of its agreement system. This is because despite the fact that these customers are located in different municipal boundaries or geopolitical areas, the utility views its system as a single entity for water quality and conservation reasons and infrastructure.

10.4   Drivers of water security

The drivers of Denver Water's strategic vision for achieving urban water security include climate change, economic demand, population growth and the political dimensions of transboundary water resources.

10.4.1   Climate change

It is projected that climate change will see an increase in statewide average annual temperatures of 2.5–5.5 degrees Fahrenheit relative to the 1971–2000 baseline. As such, typical summer temperatures in 2050 are projected to be warmer than in all but the very hottest summers in the observed record. The impacts of variation in precipitation include the following: winter precipitation is likely to increase by mid-century; spring runoff will be earlier with late summer flows likely to decrease; there will be a decrease in annual streamflow by 2050 for Colorado's major rivers; heatwaves, droughts and wildfires are likely to increase in frequency and severity due to predicted overall warming; winter precipitation events are projected to increase in frequency and magnitude; and water utilities in Colorado will be vulnerable to longer and more intense droughts, especially mega-droughts.[3] The impact of warming temperatures on Denver Water's service area is that a projected 2 degrees Fahrenheit average temperature increase would result in water supply decreasing by 7 percent due to increased evaporation, while water use could increase by 6 percent, while a 5 degree Fahrenheit increase could decrease water supply by 20 percent and increase water use by 7 percent.[4]

10.4.2 Economic demand

Water is a vital resource to the Colorado economy with the main users being residential, commercial, industrial, agricultural and mining. The Front Range, which includes Denver, represents 80–86 percent of Colorado's economy and withdraws 85 percent of the municipal and industrial water used in Colorado, while in total over 91 percent of all water withdrawn in the state is used for agricultural production.[5]

10.4.3 Population growth

Around 82 percent of the state's population lives in the Colorado Front Range. The region's population is projected to grow by 3.84 million by 2050, while the overall state's population is forecasted to increase by 60 percent by 2035 and double by 2050. Overall, Colorado projects a significant future water supply gap – over 1 000 000 acre-feet – by 2050. The gap will primarily result from increased municipal and industrial demand by 2050, with total municipal and industrial demand expected to double relative to current levels of demand. In addition to population growth, Denver itself is becoming denser with redevelopment focusing on inner-city development rather than urban expansion.[6]

10.4.4 Political

In 2012 the Colorado River Cooperative Agreement was signed in which Denver Water entered into a long-term partnership with the West Slope. Under this agreement, Denver Water's existing water rights must be used within its existing combined service area, which cannot be expanded. However, Denver Water will be able to develop an additional 10 000 acre-feet per year of water supply through conservation and reuse.

10.5 Regulatory and technological demand management tools to achieve urban water security

10.5.1 Treated water fixed charges

Denver Water charges all customers a fixed charge for water use (Table 10.3), and all types of domestic customers pay for water consumption based on an inclining block rate (Tables 10.4 and 10.5). Meanwhile, non-domestic customers have a volumetric charge for water consumption based on the seasonal water charge (Table 10.6).

Table 10.3 Fixed water charges

Fixed charges	Monthly
Service charge	$6.74

Table 10.4 Single-family treated water charges

Single-family residential	Monthly consumption (gallons)	Rate per 1000 gallons
Block 1	0–11 000	$2.75
Block 2	12 000–30 000	$5.50
Block 3	31 000–40 000	$8.25
Block 4	More than 40 000	$11.00

Table 10.5 Multifamily treated water charges

Multifamily	Monthly consumption (gallons)	Rate per 1000 gallons
Block 1	0–15 000	$3.02
Block 2	More than 15 000	$3.62

Table 10.6 All other (non-residential) treated water charges

All other (non-residential)	Monthly consumption (gallons)	Rate per 1000 gallons
Winter (Nov 1 to Apr 30)	All consumption	$1.88
Summer (May 1 to Oct 31)	All consumption	$3.76

Setting of water rates

Denver Water is not a tax-supported utility. Instead the utility relies on revenue from the sale of treated water. Denver Water's rates are set by the utility's Board of Water Commissioners. Since the utility's inception, the Board has set rates at a level sufficient to service its debt and to meet its operational and maintenance expenses. The utility's charter prohibits the utility from operating for profit: Denver Water can only charge rates that cover service costs.

Under Denver Water's charter, all of the utility's revenues go into the waterworks fund, and the money in the fund cannot be used for any purpose other than the water system. This arrangement ensures separation between City Hall and Denver Water: Denver's city government has no access to the waterworks fund, and Denver Water has no access to the city's general fund. Both funds, however, are accounted for by the city's auditor.

Denver Water has an inclined water rate structure for their residential customers. This rate structure remained unchanged for 20 years until 5 years ago when the rate structure was changed with the utility adding a fourth block. Today the utility has a pricing structure of a 4:1 ratio: the fourth block is four times the first block.

Rate structure research

Denver Water is conducting research on its rate structure to determine how water should be priced in the future. Currently, the revenue it receives is almost entirely based on usage rather than fixed fees, and when there is volatility in climate (wet/dry years), the utility faces revenue volatility. Additionally, Denver Water's pricing blocks are based on water usage in the mid-1990s; however, water demand, on a per capita basis, has declined significantly since then, resulting in the blocks being too large. Denver Water estimates that around 70 percent of its customers never leave the first block.

Rate adjustments and investing in infrastructure

Denver Water adjusts its rates by a small percentage each year rather than large increases every few years or so. Nonetheless, the rate adjustments have led to the price of water doubling over the past 10 years. Rates are increased due to the utility's large capital investment programme and the rising costs of operations. In February 2015, Denver Water's price for treated water increased by 2.2 percent to fund multi-year projects including the replacement and rehabilitation of more than 20 miles of ageing water mains throughout Denver.

10.5.2 Metering

Denver Water's customers have been fully metered since the mid-1990s with every domestic and non-domestic customer metered with automatic meter readers (AMRs). Denver Water has six water meter reading trucks that drive routes, once a month, collecting AMR meter reads. The utility bills its customers on a monthly basis: prior to 2009 billing was bimonthly. However, while townhouses are sub-metered, there is no submetering of multifamily apartments. This is an issue Denver Water will monitor as the utility needs to continue sending customers messages on how much water they use and how/what being water efficient looks like – and submetering will be essential as the city becomes denser with more multifamily buildings. Nonetheless, the cost of installing submeters is considered very expensive by the utility.

10.5.3 Reducing unaccounted-for water

Denver Water has an unaccounted-for water (UFW) rate of around 4–4.5 percent due to the utility having an active leakage detection programme. It repairs 1 percent of its infrastructure per year and is also aggressive in that if there are two leaks in the same section of pipe in the same year, the pipe is replaced. In addition, the utility's monthly billing system helps the utility reduce theft, leakages and meter reading errors. Each year Denver Water rehabilitates around 30 000–50 000 feet of pipe to protect them from corrosion and help extend the life of these pipelines.

In 2013, Denver Water began a preventative maintenance programme improving the way it replaced water mains before they become expensive. By targeting problem areas with a high number of historical breaks and by replacing ageing pressure regulating valves, the utility recorded in 2013 around 250 main breaks: a 20 percent decrease from 2012 and compared to 2011 a 37 percent decrease.

Each year a four-member Denver Water crew surveys around 500 miles of pipes searching for leaks. One of the goals of Denver Water's leak detection programme is to survey the entire distribution system – almost 3000 miles of pipes – and pinpoint leaks. The benefit of finding non-surface leaks is that it reduces expensive emergency main breaks, identifies weak pipes, reduces excavation costs and curtails water waste.

10.5.4 Protecting the quality of source water

Denver Water receives its water from snowpack and streams on US Forest Service lands. It has established a partnership with the Rocky Mountain Region of the US Forest Service, Department of Agriculture, to improve forest and watershed conditions. Through this partnership it will match the US Forest Service's $16.5 million investment to create a $33 million fund towards forest treatment and watershed projects over a 5-year period in priority watersheds critical to Denver Water's water supply. Both Denver Water and the US Forest Service have a shared interest in improving forest and watershed conditions to protect water supplies and water quality as well as habitat protection and recreational opportunities. Forest treatment – thinning, clearing and creating fuel breaks – reduces the potential for wildfires burning intensely, not only protecting lives and infrastructure but also reducing soil erosion, impacting water quality in reservoirs.

10.5.5   Water restrictions

From May 1 to October 1, customers must follow specific rules for outdoor water use that includes no watering between 10 a.m. and 6 p.m., watering to no more than 3 days per week, no watering during rain or strong winds, no watering of sidewalks and streets and repairing leaking sprinkler systems within 10 days. The utility employs a group of Water Savers to educate customers about the annual rules and best practices and even fine property owners who violate the watering rules.

10.5.6   Restrictions on alternative water supplies

Under current Colorado law, greywater may be captured and reused only in areas where the local governments have adopted an ordinance approving the use of greywater. To date, the City of Denver has not adopted such an ordinance. In addition, the capturing of rainwater is not allowed under state water law. Nonetheless, in 2009, the Colorado State Legislature passed two laws that created exemptions from the general rule. The first law states that if someone is not served by a domestic water system, for example, Denver Water, and located in designated groundwater basins or exempted collection systems, they can capture rainwater. The second law allows the state to participate in a study of 10 new developments to determine the impact of capturing rainwater on streams, rivers and tributary groundwater.

10.5.7   Rebates for promoting WaterSense-labelled products

Denver Water has partnered with the US Environmental Protection Agency for WaterSense, a national programme that allows customers to choose products that use less water without sacrificing quality or product performance. The WaterSense programme labels water-efficient products that have been independently certified for water efficiency and performance. Water-efficient products include bathroom sink faucets, high-efficiency toilets, high-efficiency showerheads, pre-rinse spray valves, landscape irrigation services and weather-based irrigation controllers.

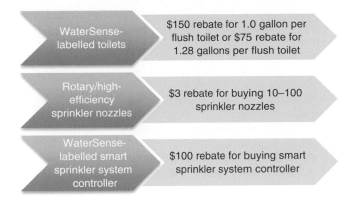

Figure 10.1 Residential rebates for water-efficient devices.

Residential rebates for water-efficient devices

Denver Water offers residential customers a range of rebates for installing the following water-efficient devices (summarised in Figure 10.1):

- *WaterSense-labelled toilets:* Customers who purchase a 1.0 gallon per flush or less per toilet installed can receive a $150 rebate, while customers who purchase 1.28 gallons per flush or less receive a $75 rebate with customers eligible for up to three toilet rebates per residence over a 10-year period.
- *Rotary/high-efficiency sprinkler nozzles:* Customers who purchase between 10 and 100 sprinkler nozzles are eligible for a one-time $3 rebate per residence over a 10-year period.
- *WaterSense-labelled smart sprinkler system controller:* Customers can receive a rebate of $100 when purchasing a WaterSense-certified smart sprinkler system controller with customers eligible for one rebate per residence over a 10-year period.

Commercial rebates for water-efficient devices

All commercial customers are eligible for rebates. Commercial customers with rebate amounts over $2500 require pre-authorisation from Denver Water. Customers applying for a rebate must agree to allow Denver Water employees to access their properties in order to verify that devices/fixtures have been installed. Commercial rebates are available for:

- *WaterSense-labelled toilets*: Commercial customers who purchase WaterSense-labelled ultra-high-efficiency toilets (1.0 gallons per flush or less) are eligible for a rebate of $150 per toilet, while customers who purchase high-efficiency toilets (1.28 gallons per flush or less) can receive a rebate of $75 per toilet installed.
- *WaterSense-labelled high-efficiency urinals*: Commercial customers who purchase a WaterSense-labelled urinal, which uses 0.5 gallons per flush, qualify for a rebate of $100 each.
- *Coin/card-operated laundry equipment rebate*: Commercial customers who replace washers with more water- and energy-efficient models can receive a rebate of $150 per washer.

- *Cooling tower conductivity controller*: Commercial customers who install conductivity controllers in their cooling towers will receive a $500 rebate per cooling tower. In addition, the cooling towers should be submetered with rebates available for submetering.
- *Submetering*: Commercial customers will receive a rebate of $40 per submeter installed in individual residential dwelling units in a multifamily building, individual commercial spaces in a commercial complex, irrigation, processes and cooling towers.
- *Commercial warewashing equipment*: Customers who purchase or lease warewashers that meet Energy Star standards for energy and water efficiency are eligible for a rebate of $300.
- *WaterSense-labelled smart irrigation controllers*: Commercial customers who purchase WaterSense-certified smart irrigation controllers are eligible for a rebate of 25 percent of material cost with multiple smart controllers allowed per account.
- *Commercial irrigation high-efficiency or rotary nozzles*: Commercial customers who purchase high-efficiency or rotary nozzles are eligible for a rebate of $3 per nozzle with a one-time rebate per property.

10.5.8   Water audits

Denver Water provides customers with a free water audit to help them become more water efficient. Customers have a range of audit types to select from.

High bill audit for single-family residential

If customers receive a water bill with an unexplained large spike in consumption, they can contact Denver Water to receive a free high bill audit to pinpoint the cause. Typically, spikes in water consumption are due to leaks both indoors and outdoors.

Large-scale irrigation systems audit

For large irrigation customers, including community associations with large common areas or commercial properties with large landscapes, free water audits can help customers identify ways to reduce outdoor water use. Customers who receive an irrigation audit will receive a report containing actual outdoor water consumption compared to annual target consumption based on weather, plant type, landscape size, map showing irrigated area and efficiency rating, potential savings estimate, issues in irrigation system encountered, scheduling recommendations and rebate and incentive programme information.

Indoor audit: Multifamily residential

A Denver Water conservation technician, along with the customer's property representative, will enter every unit for around 3–5 minutes to conduct an audit, with the utility able to complete 100 units a day. The free audit includes a free showerhead replacement, free faucet aerator replacement for kitchen and bath, inventory of leaks and potential savings if retrofits are made, historical water consumption information and rebate information.

Indoor audit: Commercial and industrial buildings

A Denver Water conservation technician, along with the customer's facility engineer or representative, will identify all water-using fixtures and processes. Based on historic consumption data, a water balance will be created, detailing how water is used throughout the facility. The free audit includes free faucet aerators, inventory of leaks, summary of savings if retrofits were made, historical water consumption information, cooling tower water-use analysis (if applicable), rebate information and incentive programme information.

Car wash certification programme

Denver Water has partnered with the Southwest Car Wash Association to develop a certification scheme for car washes in the utility's service area that have undertaken significant water-saving measures. Under this programme only certified car washes will be authorised to operate during all stages of drought response.

Challenges of promoting water audits

Denver Water often faces the problem that when the utility discusses retrofits with its customers after conducting a water audit, the customers rarely request retrofits if the pay-off period is greater than 18 months as the price of water is 'cheap'. This has been a challenge for Denver Water's conservation programme because to the utility it's a lot of water, but to the customer if they don't have an 18-month payback period, it's a deal breaker.

10.6 Communication and information demand management tools to achieve urban water security

10.6.1 School education

Denver Water has a growing education programme. There are about 10 different school systems in its service area with the utility very active in three of them. The utility wishes to be active in the remaining systems; however the service area crosses boundaries with other water utilities Denver Water supplies to. As part of its education programme, Denver Water conducts classroom tours of different components of the utility's system. The target age of Denver Water's school programme is 11–12 years old because that is when the curriculum goes through the water cycle. It takes a holistic approach to educating youth on water. The message not only focuses on the conservation part of water management but also incorporates the whole water cycle and the need to be sustainable and protect the environment, for example, do not drop or pour things down the stormwater drains.

10.6.2 Denver Metro Water Festival

In 2014, Denver Water and the One World One Water Center at Metropolitan State University of Denver hosted the first Denver Water Festival. The Water Festival is designed to provide balanced water-related education to sixth graders in Denver Water's extended service area. The festival provides children with engaging, hands-on lessons and activities to encourage students to take an active role in water conservation by providing students with the tools they need to bring wise water use to their communities. The festival also complements school curriculum on water conservation.

10.6.3 Public education and awareness: Use only what you need

Denver Water has been running the 'use only what you need' campaign over the past few years. The utility uses an advertising firm to develop and run the campaign. In the past the utility spent around $1 million a year on the campaign; however, due to more economically challenged times, the cost has been reduced to around $500–600 000. Nonetheless, the Board has been directing Denver Water to ensure that it provides the message consistently, not just in times of drought when people think about needing to conserve water but instead all the time, so that water conservation becomes a way of life.

To promote the 'use only what you need' message, Denver Water has a stripped-down car in which everything that is not required to drive is taken out. The stripped-down car is then parked at events, and people will come up and talk to Denver Water's employees about 'how weird this car is', and the employees tell them about using only what you need. The utility also has a park bench where there is only one seat with a sign on the back reading 'use only what you need'. Denver Water has also carried out a texting campaign where all the vowels out of the word 'conserve' have been removed ('cnsrv') to get the message across.

The utility tries to engage its customers by allowing them to have fun. For example, Denver Water has a toilet costume professionally made with the back of the costume reading 'stop running toilets'. The costume shows up at events and runs around and people have their picture taken with the toilet. This means the utility can engage customers on water conservation rather than handing out brochures.

10.6.4 Polling customers on water conservation

Denver Water regularly polls customers on how they feel about water conservation, and they frequently respond that they conserve water because it is the right thing to do. This aligns with Denver Water's belief that the utility should never send the message that conservation is only about reducing water bills. Customers are beginning to connect their water use with the environment they are surrounded by, for example, linking the mountains they ski on with the fresh water it provides. Denver Water will continue increasing this cultural awareness into the future, as the utility believes that water conservation is only successful if the whole state is successful.

10.6.5 Cultural change: Outdoor water use

For Denver Water the cultural shift is in how people use their water outdoors: over 40 percent of water used is for landscaping and other outdoor usage. The utility believes that there needs to be a conversation with the community on what landscapes are appropriate for Colorado's climate, for example, the region has a long history of planting bluegrass despite it not being the appropriate plant material for this type of climate due to its high watering needs. The utility believes that there is an opportunity to reduce outdoor water usage by half without changing people's lifestyles drastically. However, the utility recognises that there are revenue challenges from reducing water consumption by 20 percent.

10.6.6 Commercial partnerships to achieve cultural change in water usage

An important aspect in the cultural shift in outdoor water use involves Denver Water working with landscape and irrigation-control companies to ensure their customers – Denver Water's customers – receive the right messages on water conservation, in particular the link between green grass and higher water bills. This is difficult given that the customers are requesting their services to create green landscapes. However, low water-use landscapes can be created that require around half as much water as bluegrass. The challenge is to change people's values as to what landscape is most appropriate for the climate.

10.6.7 Targeted messaging

In 2013, Denver Water started targeting messages to customers they knew, from GIS software, were inefficient in their water use – determined by their water usage over the property's square feet area. The utility identified around 12 000 single residential customers who were using water excessively from which the utility selected 4 000 to send targeted monthly letters informing them, based on the utility's records, they are using more water than they should be and well above their neighbours' consumption.

10.6.8 Billing inserts

Denver Water believes that water bills are the best way of communicating with customers as they receive one every month. Water bills contain blurbs touting the 30-gallon target such as 'each person in an average single-family house should use roughly 30 gallons inside per day, or better yet, shoot for less!' Other billing messages include 'rethink your fixtures' and 'consult with neighbours' because 'understanding how others conserve will help you, too!' In addition, Denver Water has a neighbourhood-wide grading system that tells customers what their neighbourhood's outdoor water usage is. However, the question is whether the customers actually read the bill or just pay the amount due. As part of the rate structure study, the utility will investigate the behaviour of people reading the bill and whether they read any of the attached materials.

10.6.9 Framing water conservation messages

Instead of using the word 'conserve', Denver Water frames its messages around 'use only what you need' as the utility found that when it talked about conserving, it made people feel like they had to take up a cause. The utility also found that customers prefer the concept of 'we don't want to be wasteful' and 'don't waste – I only use what I need', and that message was successful during the economic downturn.

10.7 Case study SWOT analysis

10.7.1 Strengths

Recognising that there is cultural and political resistance to government influence, Denver Water runs an incentive-based conservation programme rather than relying on ordinances to reduce water consumption.

All customers are charged a fixed cost for water, providing the utility with revenue stability during times of lower consumption. In addition, the utility charges non-residential customers a seasonal rate to encourage water conservation during the hotter summer months.

Denver Water has an active leak detection and reduction strategy to ensure a low UFW rate. In addition to monthly billing helping customers detect leaks, the utility rehabilitates large sections of pipelines, has a preventative water mains replacement strategy and conducts annual surveys of the pipelines to detect leaks.

To lower water demand during the summer months, the utility has outdoor water restrictions in place with utility Water Savers educating customers on the rules and best practices in saving water. Water Savers even fine property owners who violate the rules.

Denver Water utilises a range of economic instruments, in addition to pricing of drinking water, to promote water efficiency and water conservation. Specifically, the utility offers residential and commercial customers a range of rebates on WaterSense-labelled devices and technologies. Furthermore, the utility provides all customers free water audits to promote water efficiency.

The utility provides water conservation messages consistently, not only during times of drought, to ensure that water conservation becomes a way of life and culturally ingrained in people.

Denver Water has started targeting customers they knew, from GIS software, were using water inefficiently. In particular, the utility sends high-consuming customers letters informing them that based on the utility's records they are using more water than they should be.

10.7.2 Weaknesses

The utility's inclining block rates for residential customers were based on water consumption rates in the mid-1990s when water consumption was much higher. Because of increased water efficiency and water conservation over time, the utility believes that the majority of customers never leave the first block, hampering further conservation efforts.

While there is universal metering of all customers, there is no submetering of multifamily apartment buildings. This hampers the utility's ability to directly link customer's water consumption with water bills. In addition, lack of mandatory submetering will pose challenges in the future as the city becomes denser with more people living in multifamily apartment buildings.

While the State of Colorado allows for greywater use in areas where the local government has adopted an ordinance approving its use, the City of Denver has not passed such an ordinance, restricting the ability of the utility to reduce pressure on scarce water resources.

The utility does not provide water-saving kits for distribution to customers, apart from devices offered through the water audit programme. In addition, the utility does not provide an online portal for customers to order water-efficient devices for installation at home following a free water audit.

While Denver Water offers commercial rebates on water-efficient devices, the utility does not offer commercial customers funding for the installation of water-efficient technologies in industrial operations that save verified amounts of water.

Denver Water often faces the problem that when it discusses retrofits with its customers after conducting a water audit, customers rarely request a retrofit as the payback period is often too long due to the price of water being 'cheap'.

10.7.3 Opportunities

Denver Water could work with the City of Denver to develop a greywater ordinance allowing large-scale non-residential users to use greywater for flushing toilets, irrigating lawns and cooling of buildings. This would reduce the amount of recycled water demanded, lowering energy costs in providing recycled water. While Denver Water provides water-saving devices during audits of multifamily apartment buildings, the utility could develop water-saving kits for free distribution at public events. To increase the uptake of water-efficient devices, the utility could enable customers to purchase WaterSense-labelled products on the utility's website with the rebate deducted from the customer's next water bill.

Denver Water could develop case studies of commercial customers who, after requesting a water audit, have made significant water savings from retrofits. These case studies could be published in industry magazines and other relevant forums to encourage more commercial customers to request a water audit.

Expanding on the car wash certification programme, the utility could approach different associations to create industry awards or certification programmes that encourage water efficiency in commercial operations.

To increase awareness on the need to save water, the utility could establish water audit programmes for children to conduct at home, irrespective of whether they live in single-family homes or multifamily apartments. Water-saving devices could then be distributed to the homes for installation, from which Denver Water could create awards for classes or schools that cumulatively have saved the most water at home over a specified period of time with prizes including equipment, such as computers, for schools.

Despite the utility's water conservation budget being lowered, there are many opportunities to harness social media in promoting water efficiency and water conservation. The utility can use Facebook competitions to promote water conservation during summer months as well as Twitter to promote individuals, businesses or neighbourhoods that have made significant water savings, particularly as the utility already provides a neighbourhood-wide grading system for outdoor water usage.

10.7.4   Threats

Climate change will impact the availability of water in Colorado and Denver with average annual temperatures expected to increase, resulting in more frequent heatwaves, droughts and wildfires along with decreased average streamflows for the state's major rivers. In addition, water is a vital resource to the Colorado economy with many users including residential, commercial and industrial sectors as well as agriculture and mining competing for scarce resources. With increased economic growth, there will likely be increased pressure on scarce water resources. Furthermore, the state's population is projected to double by mid-century, resulting in a significant future water supply gap.

10.8   Transitioning towards urban water security summary

Denver Water uses a portfolio of demand management tools to achieve urban water security (Table 10.7). However, there are numerous barriers identified by the utility in achieving further urban water security in Denver (Table 10.8).

Table 10.7 Demand management tools to achieve urban water security

Diffusion mechanisms	Tools	Description
Manipulation of utility calculations	Pricing of drinking water	All customers pay a fixed charge for water
		Residential customers pay based on inclining block rate
		Non-residential customers pay a seasonal volumetric rate
	Residential rebates	Residential rebates for toilets, sprinklers and sprinkler control systems
	Commercial rebates	Commercial rebates for toilets, urinals, washers, cooler towers, submetering, warewashers, irrigation controllers nozzles
Legal and physical coercion	Metering	All customers (domestic and non-domestic) have AMR meters
		No submetering of multifamily complexes
		Meter reading trucks collect AMR data
	UFW	Low UFW rate due to preventative leak detection programme
		Annual survey of pipes to detect leaks
		Monthly billing reduces theft, leakages and meter reading errors
	Source protection	Partnership with US Forest Service to improve watershed conditions
	Restrictions	Summertime restrictions on outdoor water use
		Denver Water deploys Water Savers teams to monitor compliance
	Alternative supplies	Recycled water provided for non-domestic customers
		Expansion of recycled water area in future
		City of Denver has not adopted greywater ordinance
Socialisation	Water-efficiency labelling	Partnership with US EPA to promote water-efficient products
	Water audits	Free water audits offered to all types of residential and non-residential customers
	Certification	Partnership with Southwest Car Wash Association to develop certification scheme
	Partnerships	Partnering with landscape companies to encourage water-efficient outdoor vegetation
Persuasion	School education	Education focuses on the whole water cycle and the need to be sustainable
		Classroom tours of Denver Water's system
		Water Festival held for children
	Public education and awareness	Consistent use only what you need campaign
		Stripped-down car to show what you only need to drive – related to use only what you need campaign
		Texting campaign to conserve water
		To make customers have fun with utility is important at public events
		Polling of customers on water conservation
		Targeted messaging using GIS
		Water conservation billing inserts
Competition	Water bills	Neighbourhood-wide grading system allows customers to compare outdoor water usage

Table 10.8 Barriers to further urban water security

Barrier	Description
Economic	The water tariff's block is too wide, meaning the majority of customers never leave the first block
	Customers rarely request retrofits as the payback period is too long due to water being 'cheap'
	Denver Water does not offer commercial customers funding for the installation of water-efficient technologies
Infrastructural/ regulatory	While there is universal metering of all customers, there is no submetering of multifamily apartment buildings
	Lack of mandatory submetering will pose challenges in the future as more people live in multifamily apartment buildings
Regulatory	The City of Denver has not passed an ordinance approving the use of greywater, despite the State of Colorado allowing it in all areas where local government has approved its use
Technological	Denver Water does not provide water-saving kits for distribution to customers, apart from devices offered through the water audit programme

Notes

1. DENVER WATER. 2015b. Water rights planning [Online]. Available: http://www.denverwater.org/SupplyPlanning/WaterRights/ (accessed 11 May 2016).
2. Ibid.
3. WESTERN WATER ASSESSMENT. 2015. Colorado Climate Change Vulnerability Study [Online]. Available: http://wwa.colorado.edu/climate/co2015vulnerability/ (accessed 11 May 2016).
4. DENVER WATER. 2015a. Climate change [Online]. Available: http://www.denverwater.org/SupplyPlanning/DroughtInformation/ClimateChange/ (accessed 11 May 2016).
5. FRONT RANGE WATER COUNCIL. 2009. Water and the Colorado economy [Online]. Available: http://www.denverwater.org/docs/assets/4bea7503-0237-e833-64a3f4c3447f588c/frwc_econ_report.pdf (accessed 11 May 2016).
6. Ibid.

11 Hamburg transitioning towards urban water security

Introduction

Hamburg's transition towards urban water security focuses on improving the sustainable management of water by reducing water consumption and treatment costs of wastewater as well as improving the utilisation of nutrients from waste for energy production. This case study analysis how Hamburg's water utility uses a portfolio of demand management tools to modify the attitudes and behaviour of water users to achieve urban water security.

The case study first provides a brief company background of Hamburg's water utility, along with an overview of the city's water supply and water consumption levels before discussing the city's strategic vision for achieving urban water security. The case study will then analyse the various demand management tools used by the city's water utility in an attempt to achieve urban water security before discussing the numerous barriers identified by the utility in achieving further urban water security in Hamburg.

11.1 Brief company background

Hamburg Wasser comprises the Hamburg Waterworks Limited (HWW) and Hamburg Public Sewage Company (HSE). Hamburg Wasser, which is publicly owned by the Free and Hanseatic City of Hamburg, has an annual turnover of around EUR 505 million.

Urban Water Security, First Edition. Robert C. Brears.
© 2017 John Wiley & Sons, Ltd. Published 2017 by John Wiley & Sons, Ltd.

11.2   Water supply and water consumption

Hamburg Wasser supplies water to two million people in the metropolitan area of Hamburg. The city abstracts its groundwater from wells between 50 metres and 350–400 metres deep and distributes water through 5500 kilometres of pipelines. The water provided to the inhabitants of Hamburg is not chlorinated; instead the water is of spring-like quality with only minimal natural treatment required to remove excess iron and manganese.

In the 1960s and 1970s, Hamburg's average water consumption level was around 160 litres per person per day. Today, Hamburg's per capita water consumption is 110 litres per person per day. The decrease in water consumption is mainly due to water meters in houses/apartments, modernisation of sanitary fittings, more efficient household appliances and consumers becoming more conscious of conserving water. In Hamburg, it is forecasted that consumption will stay static, above 100 litres per person per day over the next 10 years. Overall, Hamburg has managed to disconnect the increase in population from the consumption of water.

11.3   Strategic vision: The HAMBURG WATER Cycle

Hamburg Wasser has developed the HAMBURG WATER Cycle (HWC) concept with the objective of improving the sustainability of water management in the future. In particular, the HWC aims to reduce water consumption and treatment costs of wastewater and improve the utilisation of nutrients from waste for energy production. Overall, HWC aims to create a green, water-rich and carbon-neutral residential area that is applicable to existing cities.

11.4   Drivers of water security

The drivers of Hamburg Wasser's strategic vision for achieving urban water security include reducing the volume of imported water, climate change, population growth and rising energy costs.

11.4.1   Reducing the volume of imported water

Prior to World War II, Hamburg extracted all of its groundwater from within the city's boundaries. However, in post-war Hamburg, the city's rate of consumption increased leading to over-abstraction of groundwater supplies. This resulted in saltwater contamination of Hamburg's groundwater. While the city managed to reduce water consumption levels, the overall amount of groundwater available to the city decreased. This led to a gap between what is possible to abstract and the amount of water required by the population. Today, 75–80 percent of Hamburg's water is supplied from wells from within the city, the rest is imported from Lower Saxony and Schleswig–Holstein: Lower Saxony imposed the condition that all homes and apartments in Hamburg must have water meters installed, despite many people from Lower Saxony commuting daily to Hamburg, using the city's water supply before returning to their homes in the evening. The overall result of water metering is that Hamburg is ahead of Lower Saxony and Schleswig–Holstein in terms of water efficiency.

11.4.2   Climate change

In Hamburg the total sum of precipitation is around 760–780 ml a year. Historically, precipitation is evenly distributed over the year; however, with climate change there will be frequent heavy storm events during winter and long dry periods in the summer interrupted by sudden storm events. With storm events and climate change, Hamburg has a twofold problem: first, Hamburg is sealing its surfaces at a rate of around 60 hectares per annum. As such there are more contaminants being flushed off gardens and roads into groundwater supplies particularly during heavy precipitation events. Second, the city's dimensioning of its sewage system pipes (that collect stormwater) are still the same as if rainfall is evenly distributed over the whole year. The result is sewage pipes operating at overcapacity during heavier storm events leading to sewage water overflowing into surface water bodies. In addition, wastewater treatment plants having to treat additional storm-water during periods of heavy rainfall leads to increased energy and treatment costs (chemicals).

11.4.3   Population growth

Hamburg is one of the fastest growing cities in Germany: by 2020, the city will have grown by an additional 60 000 people.

11.4.4   Rising energy costs

Energy prices have been increasing over the past couple of years which affects Hamburg Wasser's waterworks in particular as they are 100 percent dependent on market prices for energy as they do not produce their own energy. Meanwhile, the sewage company is attempting to disconnect the development of the energy market from the treatment of water by turning waste into energy.

11.5   Regulatory and technological demand management tools to achieve urban water security

11.5.1   Pricing of water and sewage

There is a price for water and a fee for sewage: domestic and non-domestic customers pay a volumetric price of EUR 1.65 per cubic metre of potable water and a fixed tariff for water based on the water meter size (Table 11.1). For sewage, customers pay a volumetric fee of EUR 2.59 per cubic metre. Customers receive one bill with two pages – one is for the price of water, the other for the sewage fee. If a non-domestic customer uses potable water in their production process, for example, a brewery, they receive their own price for water: it is a progressive price with the actual amount dependent on the costs of connecting the user to the water supply network.

Determining the tariff level

Regarding the price of water, Hamburg's waterworks CEO provides a recommendation on what the price should be to a board of directors, which is comprised of 12 members from different stakeholders in the city, for example, a representative of the plumbing association. The board then approves or rejects the price recommendation. Meanwhile, the sewage company's CEO provides a recommendation on what the fee should be to Hamburg's Parliament, which then makes a decision on whether to accept or reject the fee for sewage. Regarding price stability, the water company has managed to keep the price of water stable for the past 5–6 years. However, the price is now being increased every 2 years

Table 11.1 Tariff for water flow

Flow	Price	Gross price
1.5 cubic metre per hour	€ 2.18	€ 2.33
1.5 cubic metre per hour of each other counter	€ 0.62	€ 0.66
2.5 cubic metre per hour	€ 5.05	€ 5.40
6.0 cubic metre per hour	€ 12.50	€ 13.38
10.0 cubic metre per hour	€ 37.30	€ 39.91
15.0 cubic metre per hour	€ 73.00	€ 78.11
40.00 cubic metre per hour	€ 86.60	€ 92.66
60.00 cubic metre per hour	€ 120.00	€ 128.40
150.0 cubic metre per hour	€ 172.70	€ 184.79
250.0 cubic metre per hour	€ 172.70	€ 184.79
Connection without a water meter	€ 74.30	€ 79.50

due to rising energy prices. Meanwhile, the fee for sewage has been increased due to salary increases of 1.5–3 percent for personal.

11.5.2   Metering

Hamburg has universal metering of all water users (domestic and non-domestic). Under the Hamburg building regulations, all apartment buildings constructed after 1987 must have submeters installed (water meters in every apartment), while in 1994 it became mandatory that all old buildings must have submeters installed retrospectively by 2004: Hamburg is the only federal state in Germany to have implemented this retrospective obligation to install submeters in old buildings.

   Each meter has to be replaced every 5 years but preferably every 3 years. Each year around 200 000 metres are replaced, a process involving water meters in houses and flats being unscrewed from the pipe and taken back to the utility's workshop where they are calibrated for reuse. Despite this labour-intensive practice, it is cheaper for Hamburg Wasser to continue using manual meters than smart meters as the cost per smart meter is too high.

11.5.3   Reducing unaccounted-for water

Hamburg Wasser has an unaccounted-for water (UFW) rate of 4 percent. To maintain a low rate, Hamburg Wasser inspects every year approximately 1000 kilometres of pipelines in the city's 5500 kilometre-long drinking water network for leaks. In addition, 20 000 valves are inspected every year to ascertain their functional efficiency. In addition to inspections, Hamburg Wasser has a long-term, comprehensive maintenance programme in which the network's old cast iron water pipes are either retrospectively lined with cement mortar or replaced with new pipes that are manufactured from ductile cast iron with PE casing and cement mortar lining. Hamburg Wasser's proactive leak management programme has reduced the damage rate in the utility's network to less than one occurrence of damage per 10 kilometres of pipeline per year.

11.5.4   Drinking water restrictions for public institutions

Since the start of the 1990s, Hamburg has ensured the city's public institutions use drinking water efficiently. This has been achieved through the development of energy and water efficiency standards that must be taken into consideration when public institutions construct new buildings or renovate existing ones. Specific water measures implemented include all sports fields must have shallow groundwater wells and the installation of approximately 7 000 waterless urinals, 65 000 continuous flow controllers in taps and 10 000 continuous flow controllers in showers, the installation of rainwater harvesting systems to flush toilets and the retrofitting of 1 500 toilets to lower flush volumes.[1]

11.5.5   Developing alternative systems: HAMBURG WATER Cycle

In Hamburg's conventional drainage system, drinking water is used to flush toilets diluting solid waste. This waste is then mixed with rainwater before entering the city's combined sewer system. Due to predicted higher volumes of wastewater (sewage and stormwater) during heavier storm events, the pipes will need to be of a larger diameter; however, larger diameter pipes have high fixed costs and long depreciation times. In particular the utility's machinery is depreciated over 12 years and the IT is depreciated over 6–7 years; however, the pipes, which accounts for 80 percent of the utility's assets, are depreciated over 70–100 years. If the utility decides to build a new water pipeline network with smaller dimension pipes, the utility will face a peak in depreciation costs – a financial loss in value. Furthermore, other assets supporting the operation and maintenance of these pipes (machinery, IT, etc.) will need to be depreciated as well in that particular year resulting in large expenses for the utility. As such, urban wastewater disposal would become more energy intensive with climate change. In addition, the current system cannot recover nutrients easily from wastewater to produce renewable energy.

The HWC is an innovative wastewater concept based on source separation. In particular, waste from toilet water is turned into energy, and greywater from washing is used for gardening, while a separate stormwater system allows rainwater to replenish ground and surface water. To test this concept, Hamburg Wasser has revitalised the former Lettow-Vorbeck military barracks into a new urban settlement called 'Jenfelder Au'. The Jenfelder Au settlement covers over 35 hectares with 770 accommodation units. The land allocation is habitation (60 percent), trade and commerce (20 percent) and green (20 percent). It is being constructed over the period 2012–2016. The overall aim of this new urban settlement is to reduce carbon emission levels to zero and utilise energy from waste. Regarding lower water consumption levels, the Jenfelder Au settlement will implement a vacuum toilet and vacuum sewage system that reduces water consumption when flushing toilets by approximately 80 percent, as suction, rather than water, will transport waste into the sewage system. In particular, the new vacuum system uses 0.5–1 litre of water per person per day for flushing compared to a conventional toilet's 6–8 litres per person per day. Therefore, waste (organic matter, nutrients) will remain concentrated rather than being diluted as happens in conventional toilets, enabling it to be extracted for anaerobic treatment before being used for energy production on-site (with zero carbon emissions).

11.5.6   Source protection and reducing energy costs

Hamburg Wasser is one of the largest estate owners in Hamburg with the company owning all the estates (forests and low-level agricultural land) around their waterworks. Hamburg Wasser has started a programme to use these estates to reduce energy costs in two ways: first, groundwater supplies can be protected from agricultural and industrial contamination, reducing energy costs of treating water. Second, Hamburg Wasser will use these spaces to build wind turbines to provide electricity for pumping water. The sewage company has been attempting to increase its production of renewable energy from biogas and wind turbines. The long-term aim is to disconnect the price of energy from the cost of treating wastewater.

11.5.7   Developing water-efficient technologies

In conventional houses and flats, Hamburg Wasser has investigated overpressure toilets, which is a toilet with a small pump operated by a battery that builds up air pressure in a small steel tank. When the toilet is flushed, the pressurised air pushes the waste out into the sewage pipe. The benefit of this type of toilet is that it can be connected to a conventional sewage system without having to modify the dimensions of the sewage pipes. In addition, the benefit of replacing conventional toilets with these overpressure toilets is an immediate decrease in water consumption by around 90 percent: from 10 to less than 1 litre per flush as only a small volume of water is required to rinse the toilet's surface.

Hamburg Wasser is attempting to determine peaks in water consumption. Over a period of a day, there is a medium consumption level (cubic metre per hour) and a peak factor. Currently, Hamburg has a large peak in the morning, a small peak during midday and a large peak in the evening when everyone comes home. However, these peak calculations are derived from historical figures that were calculated from households with older, less water-efficient showers, toilets and household appliances as compared to present-day homes. Hamburg Wasser is conducting research to determine the amounts of water consumed for specific purposes in a household. As part of this research, Hamburg Wasser has at one of its treatment plants a small area set aside to test new toilets on the market. The toilets are flushed every 5 minutes, 24 hours a day, 7 days a week (24/7) with a water meter attached to each toilet recording the amount of water used per flush. Hamburg Wasser employees have also tested new showers and water-efficient showerheads in their own homes. Hamburg Wasser also follows the latest developments in household appliances, for example, washing machines, and then calculates the potential water savings if all these new technologies and appliances are installed in a household. As such, Hamburg Wasser aims to determine the current size of these peaks for two reasons: first, the peaks determine the number of wells required to be in operation: Hamburg Wasser has 450 wells inside Hamburg and if the peak factor was found to be, for example, 1.25 not 1.4, the utility could reduce the number of wells in operation by, for example, 10 percent reducing operational costs. Second, the smaller the peak factor, the smaller the pipes need to be as the dimensions are based on cubic metres per hour. With smaller pipes maintenance costs will decrease. Furthermore, smaller dimension pipes will remove the stagnation problem common in Germany where water consumption has decreased to the point where water remains stagnant in large pipes for days at a time.

In addition, Hamburg Wasser cooperates with Melbourne's water company through staff and technology exchanges: Melbourne sent an expert on metering, while Hamburg sent an expert on energy autonomy. As part of this exchange,

Hamburg Wasser ordered three Australian meters, which can calculate water consumption rates per second, enabling the recording of consumption patterns over an entire day (showering, bathing, cooking, flushing toilets, etc.). In addition to the meters, Melbourne provided Hamburg Wasser with its own data on household water consumption the utility had collected. As part of an initial experiment, Hamburg Wasser has installed the meters in one of its departmental leaders' home to see if water consumption in Hamburg is comparable to water consumption in Melbourne. In particular, the tests aim to determine if household consumption in Melbourne and Hamburg are similar due to households in both cities using the same appliances/technologies or whether consumption rates are different, and if so why? In a more detailed experiment, Hamburg Wasser is aiming to collect data to determine, first, what the real consumption level is for the city as a whole, second, specific level of consumption for each type of consumer and household over time and, third, differences in consumption levels during weekdays compared to weekends. Overall, these figures will determine what the dimensions for Hamburg's water pipes should be in the future: if Hamburg's water consumption pattern is similar to Melbourne's, it means Hamburg could potentially use Melbourne's dimensioning for its own pipes.

11.6 Communication and information demand management tools to achieve urban water security

11.6.1 Education and awareness in schools: AQUA AGENTS

Hamburg Wasser along with the Michael Otto Foundation has created the educational programme AQUA AGENTS where students are trained to explore and discover the value and importance of water for people, nature and the economy. As AQUA AGENTS, students get an opportunity to discover the diversity of water, ask questions and find answers to water issues and develop solutions for water problems in teams. The AQUA AGENTS education programme is project based and incorporates the principles of education for sustainable development. It is tailored to the Hamburg education plan and is cross-curricular with the focus of the programme being third grade and fourth grade students. The curriculum focuses on issues of water supply, water disposal, water habitat and the port city of Hamburg. The aim is to promote an understanding of the environmental, social, cultural and economic aspects of water. As part of the AQUA AGENTS programme, classrooms become training centres with workshops focusing on case studies. Students participate in experience days where they apply their knowledge outdoors and deepen their knowledge on water. Students also acquire new knowledge of water and then pass it on to adults. In particular, students ask adults in their neighbourhoods the importance of water in their daily lives and whether they are aware of the issues such as virtual water.

11.6.2   *Public education*

Since the 1980s, Hamburg Wasser has run an extensive public awareness campaign on water conservation that aims to increase public awareness and acceptance of needing to use water carefully. To do so, Hamburg Wasser publishes a customer magazine, Wasser Magazin, which provides information, advertises open days and promotes a water forum. Hamburg Wasser hosts press conferences and exhibitions to raise awareness in addition to publishing newsletters for politicians, lobby groups, public authorities and local government officials. The utility's customer information centre also provides focused advice to customers on the careful use of drinking water and water-saving technologies. In the past Hamburg Wasser had a water-saving bus that visited markets, shopping centres, exhibitions and various events.

11.7   Case study SWOT analysis

11.7.1   *Strengths*

To promote the careful use of water, Hamburg Wasser prices its water to ensure full cost recovery of providing potable water to both domestic and non-domestic customers: the actual amount charged is regulated by a board of directors.

Hamburg Wasser has universal metering – every private home and apartment building has a water meter. In addition, every individual apartment unit has a submeter: in 1987, full universal metering was prescribed in Hamburg under the Hamburg building regulations for all new buildings, while all old buildings had to be submetered by 2004. With universal submetering it means Hamburg Wasser has the ability to directly communicate with all its water users on the need to use water carefully.

Hamburg Wasser has achieved a very low UFW of 4 percent due to the utility inspecting around 20 percent of its drinking water network per annum. In addition, the utility has a proactive pipe replacement programme where old cast iron pipes are systematically upgraded or replaced.

Hamburg Wasser is active in promoting concepts and technologies that seek to reduce water and energy consumption, in addition to carbon emissions. To reduce water usage (and energy use), Hamburg Wasser has initiated an experiment at a neighbourhood level – the HWC where a vacuum toilet and sewage system are being tested. In addition, the HWC is using greywater from washing for watering gardens and separating stormwater from the sewage system allowing the replenishment of groundwater levels, all of which reduces the amount of water requiring treatment in wastewater treatment plants. In addition, the experiment seeks to use waste from sewage into energy, further reducing energy costs.

11.7.2   Weaknesses

With the current tariff system, the utility is not financially interested in having a steep reduction in water consumption. In the next 20–50 years, with water consumption falling to around 80 litres per person per day, Hamburg Wasser will have to modify its tariff system otherwise its economic model will be obsolete: in addition to being unable to operate and maintain the water infrastructure, the utility would be unable to service its debt of EUR 1.8 billion.

For a broad range of adaptation measures to be implemented, it requires coordination and political will as the implementation will occur in existing processes of planning and decision-making which involves many stakeholders including planners, investors, politicians and even the law body, all of which are difficult to coordinate without significant political will.

The utility's current infrastructure is very inflexible in meeting new challenges, in particular managing higher volumes of stormwater in the combined sewage system. If Hamburg Wasser decides to build a new water pipeline network with smaller dimension pipes, the utility will face a peak in depreciation costs. Furthermore, other assets supporting the operation and maintenance of these pipes will need to be depreciated as well in that particular year resulting in large financial losses for the utility.

11.7.3   Opportunities

Each private home and apartment has a manual water meter that requires replacing every 3 years. As smart meters are too expensive, Hamburg Wasser could initiate a roll-out of Automatic Meter Readers (AMR) enabling customers (both domestic and non-domestic) to better understand how their behaviour affects water consumption (and their water bill).

The utility could install smart meters in specific neighbourhoods or provide businesses with them as an experiment to see how much water usage changes after installation, the purpose being to develop a business case particularly if water consumption is linked with energy usage. In addition, these experiments will provide additional data to Hamburg Wasser on appropriate pipe dimensions in the future.

Regarding smart meters, there is a potential for Hamburg Wasser to implement the use of these meters in the HWC to promote behavioural change and further reduce water consumption levels (and lower energy use and carbon emissions).

With Hamburg Wasser already conducting tests on new water-efficient toilets and showerheads, the utility could develop a water-efficient labelling scheme with the state government to promote the sale of water-efficient devices and appliances in the city. Alternatively, the utility could provide subsidies or rebates for water-efficient washing machines and low-flush toilets to reduce water treatment costs.

To further reduce the costs of providing potable water, Hamburg Wasser could promote, through various financial incentives, rainwater harvesting systems in commercial buildings, schools and other heavy users of water for flushing toilets and watering gardens. To reduce costs in the water system, the utility could in partnership with the state government provide incentives for building owners to install greywater harvesting system in all new buildings.

Hamburg Wasser is collecting data on water use by new household appliances and devices to determine peaks in daily water consumption. The purpose of this is twofold: first, by determining water use in a modern household, Hamburg Wasser can assess what diameter pipes are most appropriate in the future. With smaller pipes less water is required to move water and waste through the system, reducing energy costs of distributing and treating water and wastewater. This will also reduce carbon emissions. Second, by determining the actual size of peaks in water consumption throughout the day, Hamburg Wasser can reduce or increase where appropriate the number of wells in operation, reducing energy and carbon emissions. Hamburg Wasser nonetheless can increase its use of renewable energy with photovoltaic and wind turbines powering its water and wastewater treatment plants and well sites to decouple water consumption from energy costs and carbon emissions.

Table 11.2 Demand management tools to achieve urban water security

Diffusion mechanisms	Tools	Description
Manipulation of utility calculations	Pricing of drinking water and wastewater	Water is priced at its full economic cost of providing water services
		Customers pay for the processing of sewage
Legal and physical coercion	Metering	Hamburg has universal metering of all its domestic and non-domestic customers, including universal submetering of apartments in buildings
		All new buildings require submetering, while all old buildings were retrospectively submetered
	UFW	Hamburg Wasser has an active leak detection programme involving inspection of the water distribution network and proactive replacement of old cast iron pipes with new ones
	Alternative sources	The HWC is testing a vacuum toilet and sewage system
		The HWC project separates greywater from washing for watering plants, while stormwater is collected for replenishment of water (ground and surface)
		Hamburg Wasser is investigating water storage tanks below streets to hold excess water during periods of high rainfall for watering of trees
	Reducing treatment costs	Source protection of groundwater supplies from agricultural and industrial contamination
	Storing water during heavy storm events	Investigation of using open spaces to store rainwater during heavy storms
		Investigation of using streets to guide floodwater away from property and infrastructure
Persuasion	Education and awareness	Hamburg Wasser provides classroom materials for schools
		Public awareness is enhanced through a customer magazine, open days of the water and wastewater treatment plants and exhibitions

Table 11.3 Barriers to further urban water security

Barrier	Description
Infrastructural	Water consumption is too low resulting in water becoming stagnant in large diameter pipes
	To avoid this, water has to be flushed through the system
Economic	If there are steep reductions in water consumption levels, the current tariff structure would not be able to cover operational costs and debt servicing
	If pipes are replaced with smaller dimension pipes, there will be a peak in depreciation costs – a significant financial loss
Political	For future water systems to be implemented, it requires changes to existing processes of planning and decision-making which involves many actors

11.7.4 Threats

Hamburg Wasser has promoted the careful use of water over the past four decades in order to reduce the city's reliance on imported water. Today, the challenges of managing water sustainably in Hamburg include climate change, higher energy costs and increasing population.

The utility cannot use price flexibly to promote water conservation. Nonetheless, the price has been raised to meet rising operational costs of providing water. Meanwhile, the utility has a sewage water tariff that is regulated by the state's parliament.

With climate change there will be frequent heavy storm events during winter and long dry periods in the summer, interrupted by sudden storm events. During heavy storm events, Hamburg has a twofold problem with its sealed surfaces flushing contaminants off gardens and roads into groundwater supplies and sewage pipes operating at overcapacity leading to sewage water overflowing into surface water bodies.

11.8 Transitioning towards urban water security summary

Hamburg Wasser uses a portfolio of demand management tools to achieve urban water security (Table 11.2). However, there are numerous barriers identified by the utility in achieving further urban water security in Hamburg (Table 11.3).

Note

1. EUROPEAN COMMISSION. 2011. *European green capital award – Hamburg* [Online]. Available at http://ec.europa.eu/environment/europeangreencapital/winning-cities/2011-hamburg/index.html (accessed May 11, 2016).

12 London transitioning towards urban water security

Introduction

London's transition towards urban water security focuses on reducing total water consumption through leak detection, universal water metering of all customers and the wise use of water by all customers. This case study analyses how London's water utility uses a portfolio of demand management tools to modify the attitudes and behaviour of water users to achieve urban water security.

The case study first provides a brief company background of London's water utility, along with an overview of the city's water supply and water consumption levels before discussing the city's strategic vision for achieving urban water security. The case study will then analyse the various demand management tools used by the city's water utility in an attempt to achieve urban water security before discussing the numerous barriers identified by the utility in achieving further urban water security in London.

12.1 Brief company background

Thames Water is a privately owned water utility company. The utility was privatised in 1989 and is under the regulatory oversight of the Water Services Regulation Authority (Ofwat) – the economic regulator of the water and sewerage industry in England and Wales. Ofwat acts independently from the government and aims to provide consumers with value for money. Ofwat establishes the limit on how much individual water companies, including Thames Water, can charge their customers and aims to protect the standard of service customers receive from their suppliers.

Urban Water Security, First Edition. Robert C. Brears.
© 2017 John Wiley & Sons, Ltd. Published 2017 by John Wiley & Sons, Ltd.

12.2   Water supply and water consumption

On a daily basis, Thames Water supplies around 2.6 billion litres of tap water to 9 million customers across London and the Thames Valley through a water supply network consisting of 102 water treatment works and 87 000 miles of water mains. Thames Water also removes and treats more than 4 billion litres of sewage for 15 million customers through 350 wastewater treatment works. As such, Thames Water provides essential services to 27 percent of the United Kingdom's population.

In 2012/2013, Thames Water achieved water savings of 6.45 million litres per day versus the utility's target of 4.43 million litres per day. Currently, 70 percent of Thames Water's customers are domestic; the remaining 30 percent are non-domestic users. It is predicted that over the next 20 years domestic consumption will increase while total non-domestic consumption will decrease due to improvements in efficiency.

12.3   Strategic vision: Reducing consumption

Every 5 years, water companies in England and Wales are required to produce a Water Resources Management Plan (WRMP) that sets out how they aim to maintain water supplies. In Thames Water's 2015–2020 plan, the utility aims to, over that period, reduce the total amount of water taken from rivers by 22 million litres. In the long term, the utility aims to reach daily savings of 34.25 million litres per day by 2020. To achieve the 2020 target, Thames Water aims to manage demand through reductions in leaks, move towards universal metering of all customers and actively promote the wise use of water by all customers.

12.4   Drivers of water security

The drivers of Thames Water's strategic vision for achieving urban water security include demand outstripping supply, population growth, climate change, rising energy costs and the need to reduce carbon emissions.

12.4.1   Demand outstripping supply

Water stress is serious in Thames Water's service area with demand outstripping supply: the balance of supply and demand in London, which is three-quarters of Thames Water's customers, is currently finely balanced; however, it is forecasted that demand will outstrip supply in London by 2 percent in 2015 and 16 percent in 2040. Outside London, in the Swindon and Oxfordshire area, the deficit will be 4 percent by 2040.

12.4.2   Population growth

The population in the water supply area is projected to rise from 9 million to 10.4 million by 2040. This will increase demand for water by 230–340 million litres per day: 80 percent of this rise is expected in London.[1]

12.4.3   Climate change

By 2050, river flows in England and Wales during winter may increase by 10–15 percent and fall by 50 percent, and 80 percent in some areas, during late summer and early autumn. Overall, this could mean a drop in annual river flows of up to 15 percent. Regarding groundwater, climate change may reduce the recharge of aquifers lowering groundwater levels.[2]

12.4.4   Rising energy prices

Energy prices are predicted to rise steeply in the coming years with government forecasts suggesting an increase of around 40 percent by 2030. This will increase the costs of treating water and wastewater services from increasing energy and chemical costs.[3]

12.4.5   Reducing carbon emissions

By 2015, Thames Water aims to have lowered their greenhouse gas emissions (CO_2-equivalent) by 20 percent compared to 1990 levels. To date Thames Water has reduced their emissions by 12.5 percent, despite serving the equivalent of 3 million more customers compared to 1990 and having to meet higher water treatment standards.

12.5   Regulatory and technological demand management tools to achieve urban water security

12.5.1   Pricing of water and wastewater

Thames Water charges metered domestic and non-domestic customers a volumetric rate for both water supply and wastewater services, which is:

- Water supply: 132.48 pence per cubic metre of water
- Wastewater services: 74.82 pence per cubic metre of water

Thames Water charges a fixed tariff per annum for providing water to both domestic and non-domestic customers; the tariff is based on the size of the water

Table 12.1 Fixed charges per annum

Pipe size (millimetre)	Price (pounds)
Up to 15	29.17
20	65
25	116
30	125
40+	220

meter. All domestic customers are billed at the 15 millimetres water meter rate, while non-domestic customers are charged on the size of the meter (Table 12.1).

12.5.2 Metering

While 100 percent of Thames Water's non-domestic customers are metered, only 30 percent of the utility's domestic customers have water meters. To increase this rate, customers can request a meter free of charge. Over the period 2012–2013, Thames Water installed around 29 000 meters for customers who requested one. From 2013 onwards, Thames Water plans to commence a programme in which it will progressively meter all their customers. The programme will start in London with the utility's goal to have all domestic customers across the utility's entire supply area metered by 2030.

Smart metering

Thames Water will install more than 900 000 household 'smart' meters by 2020. This will increase the proportion of metered households to 56 percent. The smart meters will provide readings up to every 15 minutes. The smart meters will encourage customers to conserve water by charging customers on the amount they use and allow them to monitor their consumption levels. Smart meters will also help locate leaks in the utility's pipes and from those owned by customers.

Automatic meter readers

Thames Water is starting to work with a group of non-domestic customers on how automatic meter readers (AMRs) can be most beneficial in reducing water consumption rates. For this project, Thames Water has a cross section of non-domestic customer types involved including schools, universities, government buildings, council buildings, factories and office blocks. Thames Water is installing the AMRs to help these customers reduce water consumption through conventional water auditing techniques. When the customer has reached a steady state (water use is within the normal bounds for that type of business), Thames Water hopes to gain insights into how the utility should present the water data in a friendly, accurate and timely manner so customers can make consumption decisions. As Thames Water rolls this project out to more customers, it is anticipated that the software will be improved each time, driving further water savings. An additional benefit of the project is that it provides Thames Water with an insight into what other services non-domestic users may benefit from, for example, using rainwater harvesting instead of potable water for flushing toilets.

12.5.3   Reducing unaccounted-for water

Currently, Thames Water has an unaccounted-for water (UFW) rate of almost 26 percent: the leakage rate is 665 million litres per day. In the 2015–2020 plan, Thames Water will aim to reduce UFW by 59 million litres per day by 2020. As part of this, Thames Water will replace 881 kilometres of water mains in London (replacing mains with new ones where leakage is high or performance is poor). In addition, the utility has an active leak control (ALC) programme involving daily monitoring and 'find and fix' activity, pressure management (reducing pressure within the mains to extend their life and reduce leakage), and promoting customer-side leakage reduction on pipes that are the responsibility of customers through notification and subsidised repair.[4]

12.5.4   Reducing energy costs in wastewater treatment

To reduce carbon emissions, Thames Water in 2011 generated 156 gigawatt hours of renewable electricity at their operational sites, of which 151 gigawatt hours were used with the remainder being exported back into the National Grid.

12.5.5   Partnerships to install water-saving devices

Thames Water's water efficiency programme is based around the free distribution of water-saving devices such as showerheads, tap inserts and devices that physically reduce water consumption rates. For non-domestic customers with domestic water use, Thames Water offers the same water-saving devices as it does to domestic customers such as tap inserts. An added bonus of working with large non-domestic organisations is that it gives Thames Water the means of speaking to their staff directly and persuading them to have water-saving devices installed at home. Thames Water also works with schools providing them with water-efficient devices, giving the utility an opportunity to provide classroom materials for water conservation at school and home.

Thames Water is working on pilot partnership schemes with plumbers and charitable organisations to install water-saving devices in domestic customer homes. The main partnership scheme developed is with plumbers; under this scheme fully qualified plumbers are provided with water efficiency devices for installation during customer call-outs. Working off a checklist developed by

Thames Water plumbers will be paid for each device installed. While the monetary value will be low, it provides plumbers with the intangible value of credibility that comes from being associated with Thames Water. Because Thames Water will only partner with qualified plumbers, they can use this as a 'badge of honour' to increase their business. According to Thames Water, there are two main benefits the utility receives from running this scheme: first, it enables Thames Water to reach a greater number of customers as each plumber will gradually work through their communities installing water-efficient devices. Second, it is financially cheaper for Thames Water to outsource this programme as they do not have the additional costs of sending out Thames Water employees to communities installing the devices, in addition to conducting the marketing campaigns. Thames Water is also partnering with British Gas to have the utility install water-saving devices in homes they visit. Under this scheme, Thames Water compensates British Gas for each item installed. In addition, Thames Water has over 40 partnerships with local councils, charities and non-governmental organisations to install water-saving devices in people's homes. For this to succeed, Thames Water stores and distributes the devices to partner organisations for installation. However, while the quality of installations may not be as high as they could be, the quality of the engagement is high because the partners are trusted by communities. Thames Water is working with the London Wildlife Trust to design a campaign for the Trust to discuss water savings with gardeners with the utility paying the Trust to run it. The campaign will not only deliver messages the Wildlife Trust will want to say, but it will also reference water savings as being consistent with good quality gardening and that water efficiency can be a sign of a good quality garden. Thames Water is also arranging with the Wildlife Trust to sell water-saving devices, with a percentage of the sales going back to the Trust.

12.6 Communication and information demand management tools to achieve urban water security

12.6.1 Promoting water-saving devices

It is often assumed that when a customer is metered, it will induce permanent behavioural change, resulting in lower water consumption levels. However, Thames Water believes this assumption to be untrue and therefore actively promotes water-saving devices to help customers reduce their water bill. To do so, Thames Water uses various communication tools to encourage customers to request, from the utility, water-saving devices. Each communication tool has an engagement rate, the level of interaction with a customer (response rate) and a customer acquisition rate – the rate of customers having water-saving devices installed. In terms of cost effectiveness, billing inserts, despite having a low engagement rate, reach a greater number of customers than all other types of engagement tools. Therefore, in terms of volume, billing inserts have the highest acquisition rate. Regarding online engagement and acquisition, despite the utility's website only attracting a small amount of web traffic (around 200 hits per week), it has a high engagement and acquisition rate as customers are able to calculate their personal water savings if they install the devices on offer.

12.6.2   Promoting plumber visits

Thames Water favours communication tools that seek to deploy employees or representatives of the utility to customers' homes to speak directly with them. To do so, the utility sends letters to customers containing a coupon enabling the customer to redeem by phone, post or the Internet water-saving devices with the option of having a plumber install the water-saving devices for free. The reasoning behind offering free installations is the utility's research has indicated that when customers self-install water-saving devices they are frequently installed incorrectly and eventually uninstalled. Out of the three forms of redemption, customers who redeem by phone are persuaded the most to accept a plumber visit as it allows Thames Water to directly inform customers of the benefits of having the devices professionally installed. The challenge is persuading customers who redeem their coupons through the post and the Internet. People who redeem their voucher online are least likely to request a plumber visit to install the devices. A further challenge of web orders is customers redeeming their vouchers only to sell the devices online. During and after web campaigns for water-efficient devices, Thames Water detects spikes in water-saving devices being sold online.

12.6.3   Targeting demographic groups

Thames Water has discovered a trend in which those accepting plumber visits mainly fall into just two customer segments – older people and those with young families. The data found that, based on residency time, the longer a person has lived in their house, the more likely they are to contact Thames Water. In addition, when household income is factored in, those in low to middle incomes – usually households with young families and older people, retired or nearing retirement, respond the most as their water bill is a larger percentage of their income base. In order to capitalise on Thames Water's ability to effectively communicate with older customers and those with young families, the utility will be launching campaigns that target childcare centres and nurseries, as well as charities, clubs and volunteer groups that support elderly people and provide social services. Another segment of domestic customers of interest to Thames Water is vulnerable people – people on low incomes or reliant on social welfare. Thames Water believes these types of customers are least likely to request water-saving devices along with a plumber visit due to either lack of confidence or awareness. Thames Water is working with the Citizens Advice Bureau so that people, in addition to seeking general advice, will be offered a range of services including plumber visits to install water-saving devices. In addition, Citizens Advice Bureau will also distribute grants to people facing financial difficulty in paying their water bills.

12.6.4 The future: Demographic water conservation campaigns

Currently, Thames Water lacks compelling evidence of a correlation between engagement and acquisition rates and water savings. In the future Thames Water will develop, using social demographic data, targeted water conservation campaigns that focus on specific demographic groups living among a wider community. In particular, future campaigns will revolve around the distribution of demographically-tailored mail designed to encourage people of specific social groups to redeem their vouchers for water-saving devices and request a plumber visit to install these devices. To ensure acquisition rates translate into water savings, Thames Water will also develop demographically targeted conservation messages to encourage actual water savings. The overall aim is for Thames Water to gather enough data on the response rates of various forms of engagement; to be able to forecast that in a specific part of the country, containing a variety of known demographic groups, Thames Water requires X amount of money to deliver Y amounts of water savings at a specific confidence level.

12.6.5 Save Water Swindon project

The River Kennet is one of the only 200 chalk water bodies left in the world and provides a habitat for numerous aquatic wildlife including water voles and brown trout. Thames Water supplies water to 30 000 homes in Swindon from borehole sources – water that could otherwise go straight into the River Kennet to maintain a healthy flow regime. Thames Water has developed the Save Water Swindon project. This project, involving Thames Water, WWF, Swindon Borough Council and Waterwise, aims to reduce the town's reliance on the local River Kennet by saving 1 million litres per day by 2014. Swindon is a unique district metered area (DMA), where a DMA is a defined subset of the water distribution system that can be isolated by valves. The original purpose of DMAs was to allow water authorities to test the flow and pressure of water at night when a high proportion of users are inactive, and calculate the levels of leakage in each DMA. However, DMAs can also be used to calculate water conservation of residents in specific DMAs by monitoring the amount of water consumed and their response to water conservation measures. In the Save Water Swindon project, Thames Water is developing a mailing list of all its customers inside the area. From this list the socioeconomic demographics of the area can be established and the communication strategies devised for each demographic group. The overall aim is to increase acquisition rates for each demographic group. The process is not static; instead, Thames Water conducts regular customer surveys and focus groups to know how best to increase the acquisition rates. Over time, Thames Water has seen acquisition rates as a percentage of overall engagement increase from 0.5 to 8 percent over a single 12-month period – the result of Thames Water improving its ability to communicate with its customers by refining letters and even improving the quality of the envelopes. However, the use of public advertising was found to be unsuccessful with no significant increase in response rates. The utility believed this was due to poor quality adverts (created in-house) lacking creativity.

12.6.6   Education

To effectively target young people, Thames Water has established school projects on water conservation. To ensure these projects are successful at engaging young people, the utility is working with university students to learn what their perceptions are on water and what types of messages would persuade them to conserve it. Thames Water will be organising focus groups that specifically target young people in order to determine their actual level of understanding with regard to the water cycle, their values towards water and whether they want more information on water, and if so how would they like it presented. The utility will then use the results from the focus groups to determine which communication tools are most effective at 'speaking' to this particular group. At this stage the tools will revolve around social media and in particular text responses for communicating with young people.

Thames Water has gathered evidence from focus groups and surveys that the average customer lacks awareness on how water is sourced and the environmental impacts of overconsumption. To counter the lack of awareness, Thames Water has initiated the 'Rivers campaign' with the aim of informing communities, who live near the vicinity of surface water, where their water is coming from and the need to conserve it. In particular, the campaign is deliberately blunt with a picture of the local river with text stating 'this is where your water comes from'. This is followed by 'the less water you use, the less water the utility will take out of the river'. The aim of this campaign is to show how lack of water conservation can affect people's enjoyment of rivers for fishing, swimming, etc.

12.6.7   Framing of water conservation

The problem for Thames Water in promoting water conservation is that customers do not believe it is in the interests of the utility to save water: in the customers' minds something must be 'fiddled with'. Therefore, when Thames Water communicates with its customers on the need to conserve water, the utility frames the message in a way that if customers save water they will save money on both their water and energy bill. Thames Water avoids relating water conservation with the need to protect the environment because it produces a low response rate, even though it's what many customers wish the utility would say. The message that receives the largest response rate is 'the government has told Thames Water to reduce water use, therefore, the utility will tell "you" how to use less water; the benefit is you will save money'. This way the customers know why Thames Water is asking customers to save water, because the utility has been told to save.

Thames Water conducted customer research to determine which arguments are most persuasive in promoting water conservation. The research indicated Thames Water should avoid the word 'wastage' and the term 'stop wasting water' and instead use phrases that involve the terms 'responsible use' or 'use water responsibly'. The research determined that water conservation is not about wasting water or being irresponsible; instead, it is about using water responsibly because it implies there is a social good in having good quality water available rather than talking about antisocial use of water.

In 2011, Thames Water conducted a small experiment with four different letter types, all broadly similar with each saying 'use less water and save money', 'use less water and reduce your energy bill', 'use less water and protect the environment' and 'use less water because we have been told to'. The biggest response was from 'use less water because we have been told to'. This suggests Thames Water should look for government partners to act as their honest brokers. Thames Water's research also found that water conservation messages are most effective when customers are told that if they save X amount of water their energy bill will decrease by Y amount, for instance, a decrease in a customer's water bill of 20–30 pounds a year could result in a 50–60 pounds reduction in energy bills. Thames Water has also calculated the carbon equivalents of reductions in energy consumption. Nonetheless, the utility has yet to deploy either of these figures despite customers requesting the utility to do so.

12.6.8   Water audits

Thames Water provides non-domestic customers with a water audit service. However the preference is for non-domestic customers' staff to use Thames Water's online training courses to teach themselves how to conduct water audits.

12.6.9   In-house water efficiency

Thames Water has started an in-house water efficiency programme despite not being required to do so. The purpose of the programme is to develop credibility with the customer base that everyone, including the utility, needs to conserve water.

12.7   Case study SWOT analysis

12.7.1   Strengths

Thames Water has a revenue model based on fixed and variable pricing of water and wastewater services – domestic customers have a fixed cost for consuming water while non-domestic customer's fixed costs are dependent on the size of their water meter. The fixed cost component ensures future decreases in water consumption will not impact severely the ability of the utility to operate and maintain its water services.

Thames Water is aiming for universal metering with the utility taking the additional step of implementing a smart meter system over the next few years that not only promotes water conservation but enables the utility to quickly detect leaks.

Thames Water has an active water efficiency programme with the utility encouraging the installation of water-efficient devices. To ensure installation is done correctly, the utility is partnering with plumbers, other utilities and nonprofits to visit homes and install the devices properly.

The utility has identified demographic groups that are most receptive to accepting the professional installation of water efficiency devices. Using this data, Thames Water is deepening its engagement with these groups by implementing demographic-specific campaigns. At the same time Thames Water recognises there are various demographic groups that it is not reaching, and so in the future the utility will develop demographic-specific messaging campaigns.

Thames Water recognises the average customer lacks awareness on how water is sourced and the environmental impacts of overconsumption. To counter this lack of awareness, Thames Water has initiated a public awareness campaign that informs customers living near surface water on where their drinking water is coming from and how overconsumption can impact the enjoyment of these rivers.

12.7.2   Weakness

While all non-domestic customers are fully metered, less than a third of Thames Water's domestic customers are metered. This hampers water conservation efforts; nonetheless, Thames Water is aiming for universal metering of its domestic customers through a progressive metering programme, starting in London.

Thames Water needs to increase its investment in its pipeline system significantly more to achieve overall water savings as the current UFW level is very high, reducing the efficiency of the water system.

With domestic consumption predicted to increase over the next two decades, the utility does not provide a subsidy or rebate for the purchase of water-efficient toilets or household appliances.

Research indicates that Thames Water needs to develop new communication strategies as the utility is only 'speaking' to a very small part of their overall customer base. In particular, the utility is failing to target young people, particularly teenagers, who are believed to be one of the largest users of water yet have no insight on the price of water because they are not paying the water bills.

12.7.3   Opportunities

With Thames Water rolling out smart meters for its customers, the utility needs to ensure the software used is easily understood for a household to effectively save water. In particular, customers need to be able to see how their behaviour directly impacts water consumption levels. With Thames Water conducting experiments on the use of AMRs with non-domestic large users of water, there is the potential to develop alternative sources of water supply such as rainwater harvesting and greywater use, reducing the economic and environmental costs of providing potable water.

Thames Water can promote these alternative sources through rebates on the installation of water-saving technologies, with the rebate size dependent on the actual amount of water saved. This will enable Thames Water to decrease further the amount of water used by non-domestic users and reduce the gap between demand and supply in the future.

With increased metering, Thames Water could use DMAs as a tool for competition between water users. For instance, the utility could enable customers to reference their water consumption levels with other users in their own DMAs or compare their water use with other DMAs. Using smart metering and demographic data, Thames Water could also enable domestic users to compare their water consumption levels with similar households within their own or other DMAs. With regard to Thames Water's high UFW rate, increased metering will encourage customers to be more active in detecting leaks in their homes, properties and streets. In addition, smart metering of neighbourhoods will mean the utility can detect leaks in its networks quicker.

Thames Water has an active water-savings kit distribution programme where customers can redeem water-saving devices and have plumbers install these devices professionally. In addition, Thames Water could develop a water-labelling scheme to encourage customers to purchase water-efficient appliances. Alternatively, the utility could use subsidies or rebates to encourage the purchase of water-efficient appliances. To increase rates of customers requesting a plumbing visit, Thames Water can promote households who have successfully saved significant amounts of water after having plumbers install water-saving devices as points of reference for other water consumers to emulate.

With non-domestic customers, Thames Water already offers water audits for these customers. To capitalise on this, Thames Water could conduct case studies of non-domestic customers successfully saving water for other customers to emulate. In addition, by conducting water audits, the utility could provide employees with water conservation tips applicable both at work and home. It could also provide facility managers training on water efficiency.

While the utility is conducting focus groups on what messages will work with young people, the utility could increase school visits and introduce more classroom materials to educate young people on the basics of the hydrological cycle and how their actions can impact water quality and quantity. In addition, field visits by schools to Thames Water facilities can be increased so young people know where their water comes from and how it is treated.

To increase awareness of water among young people, Thames Water could organise competitions where the winners and their stories are shared through social media and other relevant outlets. Schools that have AMRs could have competitions between each other with savings from water bills being used to buy resources for classrooms such as computers. In addition, role models can be used by the utility to promote water conservation. These role models could be young people who have won social media competitions on water conservation or youth groups who have made contributions towards water conservation.

12.7.4 Threats

In London and the surrounding areas that Thames Water services, the main threats to adequate water supplies in the future are environmental, social and economic. Environmentally, the south of England has serious levels of water stress, which will be exacerbated by climate change-related variability in precipitation levels reducing surface water levels. Socially, pressure on water supplies will increase due to population growth in Southeast England, the majority of which is predicted in London. Economically, with rising energy prices predicted over the next few decades, the cost of operating the water supply system (energy and chemical costs) will increase, placing pressure on Thames Water to rely on non-price demand management tools to promote water conservation as the price of water is heavily regulated by the government.

Table 12.2 Demand management tools to achieve urban water security

Diffusion mechanisms	Tools	Description
Manipulation of utility calculations	Pricing of drinking and wastewater	The price covers the full economic cost of providing water services to customers
		Customers are charged a volumetric rate for wastewater services
Legal and physical coercion	Metering	Aim of universal metering by 2030: all non-domestic customers are metered; however, only 30 percent of domestic customers are currently metered
		Installation of smart meters for domestic users while non-domestic users are trialling AMRs
	UFW	UFW will be lowered through increased customer metering and an active leak detection programme
	Alternative sources	Through the AMR trial Thames Water will assess data to see if alternative sources of water are more appropriate for certain uses
Socialisation	Water-saving devices	Water-saving devices are distributed to non-domestic customers who use water for domestic use
	Plumber checklists	Plumbers are provided with a checklist for installing water-saving devices during visits
	Water audits	Online training courses on water auditing for non-domestic customers
	Partnerships	Partnerships with organisations to install water-saving devices
Persuasion	Public education	Focus groups to frame messages to young people better
		Public awareness on the link between drinking water and environment degradation
		Encouraging customers to redeem vouchers for water-saving devices and request a plumber to have them professionally installed
		Communication tools include phone calls, billing inserts and the Internet
		In the future communication tools will target specific demographic groups
Emulation and mimicry	Using partners to promote conservation	Wildlife Trust is running a Thames Water-designed water conservation campaign that links water conservation with being a good gardener
	Using partners to promote devices	The Wildlife Trust will sell to its members water-saving devices with the proceeds funding the Trust
	Reducing energy	Renewable energy is used to reduce energy costs
	In-house conservation	In-house conservation programme to gain credibility with customers

Table 12.3 Barriers to further urban water security

Barrier	Description
Infrastructural	While all non-domestic customers are fully metered, less than a third of Thames Water's domestic customers are metered
Financial	Thames Water needs to increase its investment in its pipeline system significantly more to achieve overall water savings as the current UFW level is very high
Economic	The utility does not provide a subsidy or rebate for the purchase of water-efficient toilets or household appliances
Demographic	The utility is only 'speaking' to a very small part of their overall customer base. The utility is failing to target young people, particularly teenagers, who are believed to be one of the largest water users
Framing	Thames Water avoids relating water conservation with the need to protect the environment because it produces a low response rate, even though it is what many customers wish the utility would say
	Research has indicated Thames Water should avoid the word 'wastage' and the term 'stop wasting water' and instead use phrases that involve the terms 'responsible use' or 'use water responsibly'
	Thames Water has calculated the carbon equivalents of reductions in energy consumption from water conservation but has yet to deploy these figures

12.8 Transitioning towards urban water security summary

Thames Water uses a portfolio of demand management tools to achieve urban water security (Table 12.2). However, there are numerous barriers identified by the utility in achieving further urban water security in London (Table 12.3).

Notes

1. THAMES WATER. 2015. Our long-term strategy 2015–2040 [Online]. Available: http://www.thameswater.co.uk/about-us/5372.htm (accessed 12 May 2016).
2. Ibid.
3. Ibid.
4. THAMES WATER. 2014. Final water resources management plan 2015–2040 [Online]. Available: http://www.thameswater.co.uk/tw/common/downloads/wrmp/WRMP14_Section_2.pdf (accessed 12 May 2016).

13 Singapore transitioning towards urban water security

Introduction

Singapore's transition towards urban water security focuses on increasing its supply of water while simultaneously reducing demand. This case study analyses how Singapore's water utility uses a portfolio of demand management tools to modify the attitudes and behaviour of water users to achieve urban water security.

The case study first provides a brief company background of Singapore's water utility, along with an overview of the city's water supply and water consumption levels before discussing the city's strategic vision for achieving urban water security. The case study will then analyse the various demand management tools used by the city's water utility in an attempt to achieve urban water security before discussing the numerous barriers identified by the utility in achieving further urban water security in Singapore.

13.1 Brief company background

Singapore's Public Utilities Board (PUB) was set up as a statutory board under the Ministry of Trade and Industry (MTI) on 1 May 1963 to coordinate the supply of electricity, piped gas and water for Singapore. In 2001, recognising that Singapore's water catchment and supply systems, drainage systems, water reclamation plants and sewerage systems are part of a comprehensive water cycle, the PUB was reconstituted to become Singapore's national water authority, overseeing the entire water cycle. The sewerage and drainage departments from the then Ministry

Urban Water Security, First Edition. Robert C. Brears.

of the Environment were transferred to the PUB. The regulation of electricity and gas industries, formerly undertaken by the PUB, was transferred to a new statutory board, the Energy Market Authority (EMA). Following this reconstitution, the Public Utilities Board is now known as PUB, the national water agency.

13.2   Water supply and water consumption

The PUB is responsible for the collection, production, distribution and reclamation of water in Singapore. The country's water supplies are known as the Four National Taps (Figure 13.1):

Local catchments Two-thirds of Singapore's land area is water catchment, and rainwater is collected and stored in 17 reservoirs around the island. Singapore is the only city in the world where urban stormwater harvesting is carried out on such a large scale.

Imported water from Johor The 1961 Water Agreement between the Johor state government and Singapore expired on 31 August 2011. Singapore continues to import water from Johor under the 1962 Water Agreement that allows the country to draw up to 250 million gallons per day (mgd) from the Johor River till 2061.

NEWater NEWater is ultraclean, high-grade reclaimed water. It was introduced in 2003 as a way to reduce the country's dependency on weather for water.

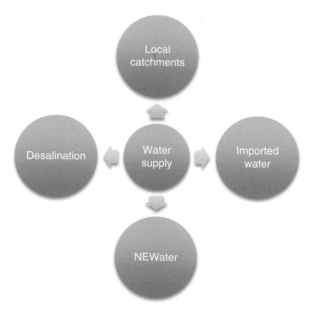

Figure 13.1 Singapore's four national taps.

NEWater is produced by purifying treated used water using advanced membrane technologies: microfiltration, reverse osmosis and ultraviolet disinfection. It has passed more than 100 000 scientific tests and exceeds the drinking water standards set by the US Environmental Protection Agency and World Health Organization. While NEWater is mainly used for industrial and air-con cooling purposes at wafer fabrication parks, industrial estates and commercial buildings, NEWater is also used to top up the country's reservoirs during dry periods. NEWater can currently meet 30 percent of Singapore's total water demand. The plan is to expand NEWater capacity so that it meets up to 55 percent of demand in the longer term.

Desalination Desalinated water has been a part of Singapore's water supply since 2005 when the country's first desalination plant was opened with a capacity of 30 mgd. The plant was the first water project to be awarded under the public–private partnership (PPP) approach. Under the contract, Singspring Pte Ltd was appointed to design, build, own and operate the plant and supply water to the PUB for a period of 20 years. A second and larger desalination plant commenced operations in 2013 under a similar PPP arrangement. The Tuaspring Desalination Plant will add another 70 mgd of desalinated water to Singapore's water supply. The plan is to grow Singapore's desalination capacity so that the Fourth National Tap will be able to meet up to 25 percent of water demand by 2060.

Water demand in Singapore is currently around 400 million mgd, with domestic users accounting for 45 percent of water use and the non-domestic sector taking up the rest. It is predicted that by 2060, the total demand for water could almost double, with the non-domestic sector accounting for about 70 percent. Currently consumption is 151 litres per person per day.

13.3   Strategic vision: Balancing supply with rising demand

Singapore's approach to water management is that while the country tries to increase its supply, the country is simultaneously reducing demand: balancing supply and demand (Table 13.1). The PUB's water conservation target is to reduce water consumption from 151 litres per person per day to 140 litres per person per day by 2030.

13.4   Drivers of water security

The drivers of the PUB's strategic vision for achieving urban water security include climate change, rising energy costs, rising population and urbanisation.

Table 13.1 Demand and supply for water resources

	Current (percent)	2030 (percent)	2060 (percent)
Demand			
Domestic	45	40	30
Non-domestic	55	60	70
Supply			
Local catchment and imported water	60	30	Approx. 20
NEWater	30	50	Up to 55
Desalinated water	10	20	Up to 25

13.4.1 Climate change

Floods and droughts as well as rising sea levels threaten the reliability of Singapore's water supply. To manage floods and more intense rainfall, Singapore is strengthening the drainage infrastructure and introducing measures to better control stormwater at source: where rain falls to the ground. Regarding droughts, the country is reviewing the adequacy of its water supply capacity and storage in case of a prolonged drought that exceeds levels never experienced before. Climate change also brings the threat of rising sea levels: Singapore is a low-lying island with much of the country only 15 metres above mean sea level, while 30 percent of the island is less than 5 metres above sea level. In anticipation of rising sea levels, the minimum reclamation levels for newly reclaimed land have been raised by 1 metre since December 2011. This is in addition to the previous level of 1.25 metres above the highest recorded tide level observed before 1991.

13.4.2 Rising energy costs

The cost of operating and maintaining Singapore's water system has increased over the past decade due to rising energy costs. With both desalination and NEWater production being energy-intensive processes, increasing their contribution to Singapore's total water supply means incurring higher energy costs.

13.4.3 Rising population and urbanisation

Currently Singapore's population is 5.3 million. It is projected that the country's population will rise to 5.8–6 million by 2020 and to 6.5–6.9 million by 2030. In addition to water infrastructure having to be built to meet increased demand, the city will become denser resulting in water pipelines competing with other services (e.g. transport, telecommunications, electrical and gas) for space even underground. The need to employ engineering techniques including tunnelling will increase the cost of implementing and maintaining the water supply network.[1]

13.5 Regulatory and technological demand management tools to achieve urban water security

13.5.1 Price of potable and used water

In the 1980s water was considered a social good and therefore priced very cheaply. In 1991, the PUB introduced the water conservation tax (WCT) to discourage excessive water use: 5 percent for domestic consumption that is more than 20 cubic metres per month and 10 percent for all water used for non-domestic uses. Between 1997 and 2000, the PUB conducted a price review, and as a result the tariff was increased to recover the full cost of production and supply. The PUB also restructured the WCT to reflect the higher cost of the next drop of water, which is essentially 'if you do not save this amount of water now, the next drop of water, which we will supply you, is going to be at a higher cost'.

Today the pricing for domestic (residential) and non-domestic (non-residential) is volume-based. For domestic users there are two tiers: for consumption rates between 1 and 40 cubic metres per month, the tariff is $1.17, which combined with the WCT of 30 percent results in a total of $1.52 per cubic metre of water consumed, while above 40 cubic metres the tariff is $1.40 with a WCT of 45 percent resulting in a total of $2.03 per cubic metre of water consumed. After the water is used, it goes through the network and is treated, and so customers pay a waterborne fee (WBF) (volume-based used water fee) and a fixed sanitary appliance fee (SAF) (fixed used water fee based on the number of sanitary appliances). These two fees are used to recover the costs of treating the used water. For non-domestic users, the water bill is based on a flat volumetric rate (no tiers) with the tariff set at $1.17 per cubic metre of water consumed plus the WCT (30 percent) leading to non-domestic users paying $1.52 per cubic metre of water. For used water, non-domestic users also pay the volume-based WBF and SAF (Table 13.2).

To help low-income families, the government has been providing grants in the form of U-Save vouchers to help offset their utility bills, including water expenses. In 2013, a household staying in one- to three-room flats received an annual U-Save voucher of $240 to $260 (or average about $20 to $22 per month), compared

Table 13.2 Pricing of water

| Tariff category | Consumption block (cubic metre per month) | Potable water | | | Used water | |
		Tariff ($ per cubic metre)	WCT (percent)	Total	WBF ($ per cubic metre)	SAF ($ per appliance)
Domestic	1–40	1.17	30	1.52	0.30	3.00
	Above 40	1.40	45	2.03	0.30	3.00
Non-domestic	All units	1.17	30	1.52	0.60	3.00

to the average water bill of less than $35 a month. This twin approach of conservation and targeted assistance ensures that all customers have access to affordable, high-quality water for the long term.

13.5.2   Metering

In Singapore, 100 percent of the PUB's domestic and non-domestic customers are metered. The PUB uses a computerised billing system incorporating a checking programme called Investigation and Report (I&R) system to verify readings taken off meters: any abnormally high or low consumption is automatically flagged by the computer during the billing process for further investigation as it could mean either leaks or theft of water.

Smart meters

The PUB will install 1200 smart meters to track water usage patterns of households and industry. Hourly data will be collected so the PUB can better understand how much water people use throughout the day. This will enable the utility to craft more effective water conservation campaigns to encourage greater water savings.

Smart water grid

The PUB has a vision of developing a smart water grid that monitors water quality and pressure and detects leakage in the water supply network by placing sensors throughout the network to collect real-time hydraulics and water quality data. Currently, the PUB has deployed 150 sensor probes and is working towards expanding the network to 300 sensor probes nationwide by the end of 2015.[2]

13.5.3   Reducing unaccounted-for water

In Singapore, unaccounted-for water (UFW) is around 4.9 percent. The majority of UFW is due to leakage. The PUB's low UFW rate is motivated by the utility's efforts to provide a high level of service to customers, in particular the reliable supply of good quality water. This means the PUB ensures customers' use is properly accounted for through accurate metering and is of a certain quality standard and that pipelines are well maintained.

Mains replacement and renewal programme in the past

Since the 1980s the installation of unlined cast iron and galvanised iron pipes has been prohibited. In addition, a mains replacement programme was implemented to replace these pipes in older parts of the network. To reduce the number of leaks from corrosion of these older pipelines, the PUB conducted an island-wide survey in 1983 to identify and replace all unlined and galvanised iron connecting pipes and unlined cast iron pipes in the water distribution system with cement mortar-lined ductile iron pipes and stainless steel or copper pipes at a cost of S$56 million. More recently, the PUB identified and replaced 280 kilometres of old, leak-prone cast iron water mains under a 5-year Mains Renewal Programme that was completed in 2004 at a cost of S$87 million.

Computerised mains replacement programme

The PUB uses a computerised system to capture information on mains including location, type, size and age of mains along with any details on previous leaks and repair work. The data is then used to plan the mains replacement programme where existing and potential problem areas are identified and prioritised for early replacement. One of the criteria used to determine if mains are due for replacement is the number of leaks occurring per kilometre per annum along sections of the water mains. Up to the end of the 1980s, a guideline of 5 leaks per kilometre per year was used. In the mid-1990s, the PUB reduced this to 3 leaks per kilometre per year. Moving forward, the PUB is reviewing the guideline to consider pipeline replacement based on 2 leaks per kilometre per year.[3]

Dynamic leak detection programme

The transmission and distribution system is composed of around 5300 kilometres of water mains. To reduce water wastage, a dynamic leak detection programme is carried out for all mains in the system throughout the year. The programme aims to trace leaks that would otherwise go unseen and undetected, especially leaks underground. Since the 1990s, the PUB uses leak noise localisers that involve visual inspection of leaks along all transmission and distribution pipeline routes and the use of leak noise localisers to quickly identify potential areas of leakage, with areas prone to leaks prioritised in this programme. The objective of the programme is to minimise the occurrence of leaks through an annual survey of the entire network with leak-prone areas surveyed two or even three times a year. To operationalise this, the entire transmission and distribution network is divided into 112 regions that are further divided into two to five sub-regions, amounting to a total of 312 sub-regions, where leak noise loggers and other detection equipment are deployed to pinpoint the location of any leaks. The PUB also monitors the dry weather flow in drains, canals and waterways to spot telltale signs of underground water leaks. During dry spells, there should be little or no water flow in the drains, canals and waterways. Substantial water flows in these waterways may indicate possible underground leaks in the vicinity.[4]

Customer relationship management

The PUB has enlisted public cooperation in the reporting of leakage, as the volume of water loss is dependent on the length of time between the occurrence of the leak and the isolation of its location. The PUB maintains a 24-hour service centre that manages all types of feedback with leakage reports directed to the Water Services and Operations branch of the centre. The service centre maintains a crew of officers and service vans to respond promptly to any water-related cases, including leakage.

13.5.4 Developing alternative water supplies

In Singapore, water supply is increased through rainwater harvesting, greywater recycling and the use of seawater for industrial purposes.

Rainwater harvesting

The construction of rainwater collection systems is regulated by the Sewerage and Drainage Act. Developers who would like to build rainwater collection systems to collect rainwater for non-potable use within their own premises can do so after an assessment of the system's design by the PUB.

Greywater recycling

Greywater means untreated used water that has not come into contact with toilet waste. It includes used water from showers, bathtubs and washbasins and water from clothes-washing and laundry tubs. It excludes used water from urinals, toilet bowls (water closets), kitchen sinks or dishwashers. The PUB defines greywater recycling as the reuse of treated greywater after the greywater has gone through treatment such as membrane filtration and disinfection to render the treated greywater safe for non-potable use. Treated greywater may be used for toilet flushing, general washing and irrigation and as cooling tower make-up water. However, use of treated greywater for high-pressure jet washing, irrigation sprinklers and general washing at markets and food establishments is not allowed as there may be public health concerns.

Use of seawater

Industries located on offshore islands or near the sea are encouraged to use seawater for cooling and process use.

13.5.5 Water Efficiency Fund

In 2007, the PUB established the Water Efficiency Fund to encourage non-domestic customers to become efficient in managing their water demand and promote water conservation within their businesses and in the community. In particular, the Water Efficiency Fund helps support non-domestic water users to implement water efficiency projects such as feasibility studies; water audits; water recycling; the use of alternative sources of water, for example, seawater in cooling processes; and even community-wide water conservation campaigns. The Water Efficiency Fund provides grants of up to 50 percent of the total cost of the water efficiency projects. Over 70 projects so far have been granted funding under this scheme.[5]

13.5.6   Water Efficiency Labelling Scheme

The voluntary Water Efficiency Labelling Scheme (WELS) was launched in 2006, the main objectives of the WELS being to keep customers well informed on their purchasing decisions and reduce water consumption by providing information on the water efficiency of products. The WELS rates products in terms of water efficiency with products receiving 0, 1, 2 or 3 'ticks': the more ticks, the more efficient the product is. To enhance the scheme, the PUB created in 2009 the Mandatory Water Efficiency Labelling Scheme (MWELS) in which all taps, urinals and flushing cisterns must display MWELS labels prominently at point of sale and packaging. Since 2011, all washing machines sold in Singapore have to carry a mandatory water efficiency label where one-tick products help consumers save 81 litres of water per wash, two-tick products can save them 102 litres of water and three-tick products can save as much as 112 litres of water. From 1 April 2014, only washing machines with at least 1-tick WELS rating will be allowed for sale and supply in Singapore. In addition, only water fittings/products that are labelled with at least a 1-tick water efficiency rating and above under MWELS can be installed and used in all new developments and existing developments undergoing renovations.

13.5.7   Water Efficient Building Certification

The Water Efficient Building (WEB) Certification (Basic) was launched in 2004 to encourage businesses, industries, schools and buildings from the non-domestic sector to include water-efficient measures in their premises and processes. Around 2500 owners or management of buildings or premises have implemented water efficiency measures with water-efficient buildings saving on average around 5 percent of their monthly water consumption.

In 2013, the PUB launched two new WEB Certification tiers: silver and gold to recognise water users who adopt the water efficiency management system and are exemplary performers in water efficiency. Recipients of the WEB Certification are certified under Industry, Building, Retail, Hotel and School Sector with certification (basic, silver, gold) valid for 3 years. The PUB has also increased its funding under the Water Efficiency Fund from 50 to 90 percent to help customers become certified under the gold/silver WEB Certification. In addition, non-domestic customers that are also implementing Water Efficiency Management Plans (WEMP) can receive 90 percent of the cost of procuring and/or installing private water meters.

13.5.8   Water Efficiency Management Plans

The WEMP was introduced in 2010 under the Ten Percent Challenge programme as a voluntary initiative to help non-domestic customers better manage and improve their efficiency in water use and help them reduce operational costs. The WEMP allows customers to understand the breakdown to water usage in their premises and develop a water balance chart; identify areas to further reduce consumption and raise efficiency; and establish an action plan that identifies measures in water savings, priorities and implement timelines. More than 370 companies have submitted voluntarily WEMP to the PUB to help them improve water efficiency. The plans help the PUB to better understand the consumption characteristics of specific industries and companies. The PUB can then use this information to set policies and propose water-saving measures for companies.

From 2015, the PUB has made it mandatory for all large non-domestic water users consuming 5000 cubic metres of water per month, or more, to submit WEMP to the PUB by June on an annual basis for 3 years. They will also be required to install private meters to measure and monitor water consumption to account for the breakdown of water use at all major water usage areas in their premises. Customers affected by the mandatory requirements as well as customers who wish to voluntarily develop their WEMP can receive funding through the PUB's Water Efficiency Fund to offset the cost of procuring and installing private water meters for the preparation of WEMP. In addition, customers can also apply to the Water Efficiency Fund to implement water-saving measures.

13.5.9 Code of Practice

The PUB has a Code of Practice for water service installations in Singapore to ensure water conservation. In all non-domestic premises, self-closing delay-action taps, which turn off even if the user forgets to turn the tap off, must be installed. In addition, dual-flush low-capacity cisterns must be installed in all new developments and existing premises under renovation.

13.5.10 Water Efficient Homes programme

In 2003, the PUB launched the Water Efficient Homes (WEH) programme that involves grassroot community groups distributing water-saving kits to all the household residents in Singapore. The kits contain flow restrictors for taps and enable customers to save up to 5 percent of monthly water consumption. Between 2003 and 2006, these water-saving kits were distributed to 910 000 households, and around 40 percent of all households had them installed. In 2007, the PUB decided to enhance the WEH by sending its employees into high-consuming homes and install water-saving devices. In addition, this enables the utility to visit households and share good practices on water conservation. The PUB also works with low-income families to install water-saving devices to help them reduce their water bill.

13.5.11 Water efficiency in new towns

The PUB works closely with other government agencies (Housing and Development Board, Urban Redevelopment Agency, etc.) to ensure government housing developments in new towns incorporate water-efficient features in the design stage: the utility can influence 80 percent of the water savings, while the owner/occupier can only influence the remaining 20 percent. Some of the water-efficient features the PUB has included in these new developments include smart water meters, use of two-tick WELS products and the facilitating of laundry water for reuse.

13.6 Communication and information demand management tools to achieve urban water security

13.6.1 School programmes: Time to Save water

The PUB has worked with the Ministry of Education to incorporate water conservation topics into the social studies syllabus for Primary 3 students. At the same time, all Primary 3 students participate in the 'Time to Save' programme that involves participants using a timer and activity booklet to track their shower timings for a week. During the week, the students will also become junior water advocates encouraging their family members to take shorter showers and sharing water conservation tips with their neighbours. As part of this programme, the PUB's mascot 'Water Wally' will perform during assembly at primary schools. The performance includes a 5-minute 'shower dance' to demonstrate how water conservation can be simple and fun.[6]

13.6.2 Public education: Fostering the emergence of a water-saving culture

When the PUB analyses the breakdown of consumption in a typical household, the utility notes that, while a lot of water savings is due to technological solutions – hardware, water-efficient fixtures and so on – more water can be saved if people change their attitude and behaviour towards water. The PUB has conducted demographic studies and found there are certain groups of consumers in Singapore with regard to water conservation. There are 'very devoted conservationists', the 'non-action believers' and 'non-conservationists'. The aim of the PUB's water conservation programmes is to push non-action believers and non-conservationists into being devoted conservationists.

13.6.3 Water Volunteer Group programme

In 2006, the PUB launched the Water Volunteer Group (WVG) programme in which grassroot volunteers, with the help of PUB officers, go door to door to educate households on behavioural aspects of water conservation. In particular, the WVG challenges residents to save 10 litres per person per day.

13.6.4   Water Conservation Awareness Programme

The Water Conservation Awareness Programme was launched in 2011 to raise awareness on the importance of using water wisely. The programme consists of three initiatives: a revamped water-saving kit with increased options for water thimble sizes and stickers, a water audit project in which students conduct water audits and reduce water usage at home and water conservation training by maid agencies with a DVD and handbook in English and Bahasa Indonesian for domestic helpers. The Awareness Programme also conducts roadshows at community events with exhibits to better illustrate how the public can save water as well as television commercials aimed at increasing public awareness of simple water-saving habits and installation of thimbles.

13.6.5   Ten Percent Challenge for non-domestic customers

The Ten Percent Challenge aims to challenge the non-domestic sector to save 10 percent of their monthly water consumption. To achieve a 10 percent reduction in water consumption, the PUB provides building owners and managers with a website where they can learn about water efficiency. In addition, the PUB has developed in cooperation with Singapore Polytechnic a Water Efficiency Manager (WEM) course to equip facilities managers with the knowledge and skills to conduct water audits and apply water efficiency measures to reduce water consumption in commercial/residential buildings. The target audience of the WEM course is facilities and estate managers, building owners and engineers and architects.

13.6.6   Watermark Award

The Watermark Award was introduced in 2007 to recognise individuals and organisations for outstanding contributions and commitment to protecting and raising awareness of water resources. Recipients of the award are role models who provide inspiration to all water users to take ownership of the water and play an active part in ensuring Singapore's water sustainability.[7]

13.6.7   Water efficiency certificates for building owners

In the non-domestic sector, a certain portion of water usage is domestic in nature (toilet flushing, showers, etc.). To reduce this consumption, the PUB recognises building owners who, in addition to meeting regulatory requirements, voluntarily reduce water consumption levels in their buildings by certifying their premises as being water-efficient. The PUB also works closely with the Building and Construction Authority to ensure newly constructed buildings have water efficiency measures in place.

13.7 Case study SWOT analysis

13.7.1 Strengths

The PUB charges domestic and non-domestic users a WCT to discourage excessive water consumption. Customers also pay a WBF (a volume-based used water fee) and a fixed SAF (fixed used water fee based on the number of sanitary appliances) to recover the costs of treating the used water.

All customers – domestic and non-domestic – are fully metered in Singapore enabling the utility to deploy a computerised billing system that automatically detects abnormalities in the water distribution system that could indicate theft or leakage.

To maintain a low UFW rate, the PUB has a computerised mains replacement system that identifies existing and potential problem areas for early replacement, in addition to a dynamic leak detection programme that each year surveys the entire system, in some places more than once, and a 24-hour service centre that customers can report leaks to.

To promote water-efficient devices and appliances, the utility has a dual voluntary and MWELS so customers are informed on purchasing decisions. In addition, to encourage water efficiency in buildings, the PUB has developed a WEB Certification.

To reduce non-domestic water consumption, the PUB has introduced WEMP for customers to understand water usage in their premises as well as identify areas to reduce consumption and increase efficiency. This scheme became mandatory for all large-scale customers requiring the submission of annual water plans as well as the installation of private meters at all major water-using areas.

To promote behavioural change among domestic customers, the PUB has established volunteer groups that work with PUB officers to educate households on water conservation. The PUB has also established an awareness programme that provides water-saving kits as well as a water audit checklist for students to conduct audits at home.

The PUB recognises outstanding contributions and commitments towards protecting and raising awareness of water resources through the Watermark Awards. Recipients are considered role models who inspire others to take ownership of water resources and ensure the long-term sustainability of water.

13.7.2 Weaknesses

The PUB maintains a mainly large-scale technological focus on reducing water consumption for both domestic and non-domestic customers. While technology can increase efficiency in household water consumption and industrial operations, it must be recognised that the changing of attitudes and behaviour towards water can make significant water savings.

The PUB does not maintain a wide range of educational programmes for both schools and the public to increase the emotional message of needing to save water.

The PUB's classroom programme only targets young children on water conservation. The utility has not developed curriculum material to educate teenagers and young adults on in-depth water-related topics such as the hydrological cycle. While the utility organises roadshows, the utility does not hold year-round education workshops on specific water-related topics or partner with nonprofits to spread targeted messages.

The utility does not offer rebates for customers to purchase water-efficient devices and household appliances. In addition, the utility does not offer non-domestic customers rebates on the purchase of water-efficient devices and appliances for domestic uses of water.

The PUB understands there are several customer segments in Singapore with differing attitudes and behaviour towards the environment and water; however the utility does not conduct targeted water conservation messaging.

13.7.3   Opportunities

The PUB should host community events to encourage changes in the attitudes and behaviour of all customer segments towards scarce water resources. The utility in particular could hold water days that involve schools, community groups and nonprofits demonstrating to the public ways to conserve water.

The Watermark Award could have categories for domestic as well as institutional customers such as community groups, nonprofits and schools that have made outstanding contributions towards water conservation. In addition, a Watermark Award could be awarded to an outstanding citizen-of-the-year who has made significant contributions towards the sustainable management of water.

The utility could compile case studies of companies receiving Watermark Awards to encourage water efficiency in their competitors and respective industries. In addition, case studies of non-domestic customers using water efficiency funding to make significant savings can be publicised in newspapers, magazines and so on and on the PUB's website to encourage more companies to seek funding.

To encourage further water conservation by non-domestic customers, the PUB could create Watermark Industry Awards to recognise businesses in specific industries that have made outstanding contributions towards protecting water resources.

To encourage community-wide water conservation, and foster a competitive spirit in reducing water usage, the PUB could provide a water consumption grading system for all customers residing in the 312 sub-regions the utility has created for managing its leak detection programme. Each sub-region could receive a grade for their water usage and compare that with an overall average.

To reduce rising energy costs of providing water in the future from desalination and NEWater, the PUB can deploy smart meters to target specific water users and groups to encourage water conservation and energy savings. For instance, smart meters could be made mandatory for all large non-domestic users of water. The mainstreaming of smart meters will also enable the PUB to determine the optimal operational level of the water supply network to meet peak demand during

Table 13.3 Demand management tools to achieve urban water security

Diffusion mechanisms	Tools	Description
Manipulation of utility calculations	Pricing of drinking and wastewater	Domestic customers are charged a fixed and volumetric tariff for drinking water and non-domestic customers a flat volumetric tariff for drinking water
		All customers pay a water conservation tax on drinking water and a volumetric waterborne and fixed sanitary appliance fee for treating used water
	Funding for water efficiency	Water Efficiency Fund (WEF) encourages non-domestic users to become efficient in water usage
		WEF available for community-wide conservation campaigns
Legal and physical coercion	Metering	All domestic and non-domestic customers are metered
		Abnormal water consumption is flagged for investigation
		Smart meters will be trialled to track domestic usage patterns
	UFW	Island-wide water mains replacement and renewable programme
		Computerised mains renewable programme is used to plan the replacement of water mains
		The entire water mains network is surveyed each year
	Restrictions	Only water fittings and products with mandatory water efficiency labels can be installed in new developments/existing ones undergoing renovations
		Only washing machines with at least one-tick WELS rating allowed for sale and supply
		Code of practice mandates installation of water-efficient devices in new developments/existing ones undergoing renovation
Socialisation	Water-saving devices	Water Efficient Homes programme involves community groups distributing kits to households
	Water efficiency labelling	Voluntary Water Efficiency Labelling Scheme to inform consumers on purchasing decisions and mandatory labelling for specific devices and appliances
	Water-efficient building certification	Water Efficient Building Certification encourages non-domestic customers to be water-efficient
	Water efficiency management plans	Low-volume non-domestic customers submit Water Efficiency Management Plans voluntarily
		All large-volume non-domestic customers must submit annually Water Efficiency Management Plans
	Water efficiency in new towns	The PUB cooperates with other government agencies to ensure government housing developments install water-efficient devices and appliances plus smart meters and laundry water reuse
	Water efficiency course	The PUB has developed with a local polytechnic Water Efficiency Manager course for building managers
Persuasion	School education	Time to Save water programme in schools
		The PUB's mascot 'Water Wally' performs at primary schools
	Public awareness	Distribution of water-saving kits, water audits, water conservation education for maids and roadshows at community events
		TV commercials on water-saving tips and how to install water-saving devices
		Ten Percent Challenge for non-domestic customers
Emulation and mimicry	Classroom water heroes	Students in the Time to Save water programme become junior water advocates who share conservation tips with neighbours
	Water volunteers	Water volunteer groups go door to door educating households on behavioural aspects of water conservation
	Water awards	The PUB recognises individuals and organisations who have made outstanding contributions to protecting water and raising awareness
	Water efficiency certificates for building owners	The PUB recognises building owners who, in addition to meeting regulatory requirements, voluntarily reduce water consumption levels in their buildings by certifying their premises as being water-efficient

different times of the day, for example, less pumping will be required during the early hours of the morning when the majority of the population is asleep. This in turn will reduce operational and maintenance costs.

13.7.4 Threats

Climate change will threaten the reliability of Singapore's water supply with both floods and droughts impacting infrastructure and availability of supply. To manage urban floods, the PUB is strengthening the drainage infrastructure and introducing measures to better control stormwater at source: where rain falls to the ground. Regarding droughts, the country is reviewing the adequacy of its water supply capacity and storage in case of a prolonged drought that exceeds levels never experienced before. Rising sea levels will also impact the availability of water supply from damage to infrastructure.

With the PUB relying on NEWater and desalination to bridge the gap between demand and supply as well as reduce the amount of imported water, the costs of operating and maintaining the water system will increase over time due to rising energy costs. This will place pressure on the utility to increase the price of water, impacting household affordability and the competiveness of businesses.

Singapore's population is projected to rapidly increase over the next couple of decades. In addition to water infrastructure having to be built to meet increased demand, the city will become denser resulting in water pipelines competing with other services. This will increase the cost of implementing and maintaining the water supply network, potentially leading to the price of water increasing even if consumption decreases.

Table 13.4 Barriers to further urban water security

Barrier	Description
Economic	The PUB does not offer rebates for domestic customers to purchase water-efficient devices and household appliances
	The PUB does not offer non-domestic customers rebates on the purchase of water-efficient devices and appliances for domestic uses of water
Institutional	The PUB does not maintain a wide range of educational programmes for both schools and the public to increase the emotional message of needing to save water
	With schools the utility only targets young children on water conservation. The utility has not developed curriculum material to educate teenagers and young adults on in-depth water-related topics
	The PUB does not hold year-round education workshops on specific water-related topics or partner with nonprofits to spread targeted messages
Technological	The PUB maintains a mainly large-scale technological focus on reducing water consumption for both domestic and non-domestic customers. However, it is the changing of attitudes and behaviour towards water that can make significant water savings
Demographic	The PUB understands there are several customer segments with regard to their attitudes and behaviour towards the environment and water; however the utility does not conduct targeted water conservation messaging

13.8   Transitioning towards urban water security summary

The PUB uses a portfolio of demand management tools to achieve urban water security (Table 13.3). However, there are numerous barriers identified by the utility in achieving further urban water security in Singapore (Table 13.4).

Notes

1. PUB. 2015. Our water, our future [Online]. Available: http://www.pub.gov.sg/mpublications/OurWaterOurFuture/Pages/default.aspx (accessed 18 May 2016).
2. Ibid.
3. PUB. 2011. Low unaccounted for water [Online]. Available: http://www.pub.gov.sg/general/watersupply/Documents/UFW_Guidebook.pdf (accessed 18 May 2016).
4. Ibid.
5. MEWR. 2015. Sustainable Singapore blueprint [Online]. Available: http://www.mewr.gov.sg/ssb/ (accessed 18 May 2016).
6. PUB. 2015. Our water, our future [Online]. Available: http://www.pub.gov.sg/mpublications/OurWaterOurFuture/Pages/default.aspx (accessed 18 May 2016).
7. Ibid.

14 Toronto transitioning towards urban water security

Introduction

Toronto's transition towards urban water security focuses on reducing peak day water demand and reducing wastewater flows with significant economic, environmental and social benefits. This case study analyses how Toronto's water utility uses a portfolio of demand management tools to modify the attitudes and behaviour of water users to achieve urban water security.

The case study first provides a brief company background of Toronto's water utility, along with an overview of the city's water supply and water consumption levels before discussing the city's strategic vision for achieving urban water security. The case study will then analyse the various demand management tools used by the city's water utility in an attempt to achieve urban water security before discussing the numerous barriers identified by the utility in achieving further urban water security in Toronto.

14.1 Brief company background

Toronto Water serves 3.4 million residents and businesses in Toronto, as well as portions of York and Peel. Under the Municipal Act, the city of Toronto owns the water system with Toronto Water managing it. Toronto Water's mission statement is to provide quality water services through supplying drinking water and the treatment of wastewater and stormwater to residents, businesses and visitors in order to protect public health, safety and property in an environmentally and

Urban Water Security, First Edition. Robert C. Brears.
© 2017 John Wiley & Sons, Ltd. Published 2017 by John Wiley & Sons, Ltd.

fiscally responsible manner. The utility's vision is to be a leader in achieving excellence and efficiency in all aspects of water service delivery.

Toronto's Water Supply Bylaw, enacted in 2008, harmonises practices and procedures and standardises fees ensuring Toronto Water provides the same, or better, levels of service to all its customers across the city, without imposing any new service fees. Specific measures include providing reliable metering information, improving operational efficiency and protecting the integrity of the water supply system.

14.2 Water supply and water consumption

Toronto Water maintains a $28.2 billion water and wastewater system, $9.1 billion of which is the water system comprising four water filtration plants, 11 reservoirs and four elevated storage tanks, 5501 kilometres of distribution mains and 18 pumping stations.

Toronto is built on the side of a long, sloping hill and so providing water to the community requires pumping. Toronto Water uses pumps to raise water pressure and push the water from the lake level to the higher elevated areas. To ensure adequate water pressure across the city, the entire Toronto area is divided into six levels or pressure zones with each zone divided into pressure districts, with each zone selected based on the ground elevation range of its particular area of Toronto. Because water needs to be pumped one to three zones upwards, Toronto Water has 18 pumping stations located in different pressure districts. To meet high water demand and ensure adequate pressure, the utility also has ground-level reservoirs and elevated tanks to meet peak demand while providing stable water pressure.

Toronto Water produces potable drinking water by treating and cleaning raw water from Lake Ontario. In particular, water intake pipes extend into Lake Ontario and collect raw water. In some parts of the city, the intake pipes extend as far as 5 kilometres offshore. The water then passes through travelling screens that remove large objects and debris at the entrance of the plant. Chlorine is added as well as alum, which causes small particles to clump together to form larger groups of particles − flocs. In the settling basin, heavy flocs sink to the bottom with cleaner water at the surface drawn off through spillways leading to filtering basins where it passes through filters made of graded gravel, fine sand and carbon. After filtration the water goes into holding basins where chlorine and fluoride are added. The last step in treatment is the adding of ammonia to stabilise the chlorine ensuring drinking water is safe.[1]

There are three main categories of consumers in Toronto: single-family residential customers in single-detached, semi-detached, row housing (3–6 units) or plexes (2–6 units); multi-unit residential in high-rise or low-rise apartment buildings, condominiums or cooperatives, each with greater than six units; and industrial, commercial and institutional (ICI) that include offices, retail outlets, hotels, hospitals, factories, warehousing, manufacturing, government buildings and schools. The water demand per sector is in Table 14.1.

Table 14.1 Demand for water per customer category

Customer category	Water demand (percent)
Single-family residential	34
Multiunit residential	19
ICI	33
Non-revenue water	14

14.3   Strategic vision: Toronto's Water Efficiency Plan

Toronto's Water Efficiency Plan called for the city to reduce peak day demands and reduce wastewater flows. Building on the Municipality of Metropolitan Toronto's 1993 target of reducing water consumption by 15 percent by 2011, Toronto set the goal of reducing peak day demand by 275 million litres per day and wastewater flow by 86 million litres per day. In addition to saving infrastructure costs, the Water Efficiency Plan listed additional benefits the city would see including avoidance of energy and chemical costs, reductions in carbon emissions, improvements to surface water quality, savings in water bills and preserving water for future generations.[2]

14.4   Drivers of water security

The drivers of Toronto Water's strategic vision for achieving urban water security include in the past ensuring the city meets water conservation targets to avoid infrastructure upgrades and, today, using water efficiently while managing declining revenue streams.

14.4.1   Previously: Meeting specific water conservation targets

The Water Efficiency Plan projected that to meet projected population growth of more than a quarter million people – from 2.59 million in 2001 to 2.86 million in 2011 – and an increase in employment – from around 1.45 million jobs in 2001 to 1.62 million in 2011 – the city could either increase the system's capacity by expanding water and wastewater infrastructure at a cost of $220 million or consider implementing water-efficient measures to help 'free up' capacity within the existing system.

14.4.2   Today: Using water efficiently

Since the publication of Toronto Water's Water Efficiency Plan, Toronto has experienced a downward trend in water consumption over the last decade despite increased population growth. Over the past 7 years, there has been a decline in base water consumption (over the period of October to April) by 2.1 percent annually on average, while summer consumption has reduced over the same time period by 1.7 percent annually. As a result of lowered consumption forecasts, Toronto Water's 2012–2021 Capital Plan was reduced by $1.132 billion. A further decline of 1 percent (compared to 2012 actuals) in water consumption is projected for 2014. While this does indicate that the decline in water consumption is beginning to level off, it does continue to further reduce revenues available to fund the capital programme.[3] At the same time, the utility is faced with ageing infrastructure with a $1.6 billion backlog ($1 billion for underground assets) of maintenance and upgrades required. This is a challenge given the utility's 10-year Capital Plan relies primarily on successive water rate increases to fund continued infrastructure investment and conform as a pay-as-you-go financing strategy. In addition, Toronto Water faces increased legislative and regulatory reform impacting both operating and capital budgets. Specifically, the provincial regulations include an expansion of the Safe Drinking Water Act requiring municipalities to publish annual reports describing the operation of the water system and the results of testing required to ensure residents are provided with safe drinking water and a new bill for the Clean Water Act which will provide protection for municipal drinking water supplies through developing collaborative, locally driven, science-based protection plans by municipalities, conservation authorities and the public. As such, there has been a change in focus on how the utility will manage water sustainably. In particular, the utility is moving from meeting specific targets to using marketing and promotions to ensure the general public understands they have to be efficient in using water.

14.5 Regulatory and technological demand management tools to achieve urban water security

14.5.1 Water rate for water, stormwater and sewer

Toronto Water is rate supported – the utility does not rely on the property tax base to support its operating and capital budgets. The utility has a single water rate for domestic and ICI customers who consume less than 6000 cubic metres. ICI customers that consume over 6000 cubic metres can apply for an industrial water rate. The water rate is an all-inclusive rate that includes water supply, stormwater management and sewer fees (Table 14.2).

Setting of the water rate

The water rate is based on econometric models which contain variables including forecasted consumption based on historical data, events that are likely to increase or decrease water consumption and operating budget and capital investment

Table 14.2 Metric water rates

Metric water rates	Rate if paid on or before due date (per cubic metre)	Rate if paid after due date (per cubic metre)
General water rate Block 1 – Applied to all water consumption, including the first 6000 cubic metres	$3.1945	$3.3626
Industrial water rate Block 2 – Applied to water consumption over 6000 cubic metres for businesses participating in the Industrial Water Rate programme	$2.2361	$2.3537

requirements for the following year. The outcome of the model is the overall funding required to operate and maintain the water system that in turn sets the rate for water.

Beginning in 2006, Toronto Water implemented a planned multiyear water and wastewater annual rate increase of 9 percent per year for 9 years. The annual multiyear rate increase strategy was planned to generate revenues required to fund Toronto Water's operations and balance infrastructural renewal needs with new service improvement while meeting new regulatory requirements. The 9 percent rate increase was in response to a projected increase in population that did not eventuate – with lower than projected consumption, the utility's revenue decreased, reducing the amount of funding available for capital infrastructure investments and operating budgets. The result is the utility having to continue raising rates over time to meet the revenue shortfall.

In 2015, Toronto Water increased rates for water and wastewater by 8 percent over the period of 2015–2017 in order to reinstate around $1 billion in capital funding – lost from declining water consumption rates – needed to fund infrastructure projects as part of the $11.040 billion 2015–2024 Capital Plan.[4]

14.5.2   Metering

Toronto's Water Supply Bylaw mandates the installation of water meters at all properties. Toronto Water is working towards universal metering with the utility implementing a 5-year plan to have all customers metered (Table 14.3). Starting with customers without meters and paying a flat rate, the utility, through a subcontractor Neptune Technologies, will install automatic meter readers (AMRs) for free ward by ward. In addition, all customers with existing manual meters will have them replaced with new AMRs to ensure the metering system is efficient with up-to-date technology. According to Toronto Water, the benefits of the new system include keeping better track of water consumption across the city, detecting water loss more quickly and eliminating the need for utility

Table 14.3 Progress towards universal metering

Metering		2012	2013	2014	2015	2016	2017	2018
Percentage of AMRs	Target	36	57	83	95	96	97	98
	Actual	33	62	93	n/a	n/a	n/a	n/a

CITY OF TORONTO. 2015c. Service level review public works and infrastructure committee presentation [Online]. Available: http://www.toronto.ca/legdocs/mmis/2015/bu/bgrd/backgroundfile-85365.pdf (accessed 12 May 2016)

personnel to manually read meters. Eventually, Toronto Water's customers will be able to access an online portal to view their consumption and potentially compare that against a specific average, for example, the average of their block, ward or postal code.

Prior to the installation of AMRs, Toronto Water sends out letters informing customers of their installation along with information on how to contact the utility if they have any questions. When customers receive their first AMR-based water bill, the utility will also provide water efficiency billing inserts to help customers reduce their consumption levels and reduce water costs, for example, check the toilets for leaks if they see unusually high readings. Previously, customers received water bills four times per year based on two meter reads and two estimates. Now with AMRs, every water bill sent out four times per year is based on actual reads.

Reclassifying customers

Currently, large apartment buildings may be classified as an ICI or residential user of water. As Toronto Water installs new AMRs, the utility will ensure each building is classified correctly, ensuring the portion of domestic, commercial, institutional users of water will be more accurate in the next couple of years.

14.5.3 Reducing unaccounted-for water

In 2011, the city of Toronto authorised Toronto Water to implement a citywide Water Loss Reduction and Leak Detection programme. This programme was based on a detailed Water Loss Assessment and Leak Detection Study that sought to quantify water losses and unbilled authorised consumption, that is, non-revenue water, including water distribution leakage, loss of water through water main breaks, use of water for firefighting purposes, operations and maintenance of the distribution system including hydrant flushing and unmetered consumption, for example, irrigation systems at some city parks and facilities. The study found water losses were around 8–10 percent of production total, corresponding to an estimated annual value of $30 million in treatment and transmission costs.

Table 14.4 Water main breaks per 100 kilometres of water distribution

Breaks		2012	2013	2014	2015	2016	2017	2018
Water main	Target	20.8	20.8	20.8	20.8	24.8	23.1	23.1
breaks	Actual	18.2	25.1	29.6	n/a	n/a	n/a	n/a

Water main replacement and rehabilitation

Toronto Water replaces approximately 40–60 kilometres of water mains each year. Priorities for water main replacements are determined using a combination of age, break frequency, material, operational requests, hydraulic performance and future growth and to minimise cost and disruption to the local community in coordination with other construction programmes including road, gas, sewage and so on. In 2014, Toronto Water invested $83 million as part of the water main replacement and rehabilitation programme.[5] In addition, Toronto Water's budget included financing of $0.427 million to establish a water loss reduction and leak detection team. The team will consist of a project lead, engineering technician, technologist, and water maintenance workers.

Over the period of 2012–2014, there was a rising trend in water main breaks due to severe cold weather fluctuations and ageing water mains. Over the period of 2016–2017, the plan is to maintain water main break and repair levels of typical climate years (Table 14.4).

14.5.4 Capacity Buy Back programme

Toronto Water is focusing on increasing water efficiency of ICI customers with the view that not only does it reduce their operating costs, and increase their competitiveness, but it also ensures they do not waste water unnecessarily. Specifically, Toronto Water has calculated that water efficiency improvements by ICI users could reduce the need to build new water and wastewater treatment plants that would cost $2.5 billion over the next 20 years.

Toronto Water's Capacity Buy Back programme encourages and rewards ICI organisations that reduce water use. Under this programme, Toronto Water buys back water capacity that has been freed up by participants who have reduced water use in their operations. By implementing permanent process or equipment changes that save water, ICI organisations are eligible for a cash rebate. Specifically, the programme helps identify how ICI customers can reduce water use, offers a one-time case rebate of up to 30 cents per litre of water saved per average day and helps participants save money over the long term through reduced water bills. To apply for this programme, ICI customers need to register their interest with Toronto Water. Toronto Water will then arrange for a water audit of the organisation's premises by a professional engineer. Following the visit, the engineer will develop a detailed water audit report that includes a list of eligible process or equipment changes the organisation would have to make and an estimation of savings made. Once an ICI customer has implemented these changes and had them verified by Toronto Water, the organisation will receive a cheque for the amount of water saved.

14.5.5   Industrial Water Rate programme

Toronto Water has an industrial water rate to support the growth of businesses using water for processing purposes and to encourage water conservation. To qualify for the industrial water rate, customers have to consume more than 6000 cubic metres of water annually, fall within the industrial property tax class, be in full compliance with the city's Sewers Bylaw and submit a comprehensive water conservation plan to Toronto Water.

The water conservation plan needs to identify all uses of water in the plant and identify all opportunities as well as an estimated payback period. Toronto Water uses independent consultants to review the plan and check to see if there are any more opportunities available to conserve water, which may have to be implemented if the customer is to receive the rebate: any aspect that has a payback period of 5 years or less has to be implemented. Toronto Water also makes annual checks on the progress of the water conservation plan to ensure compliance with the sewer bylaw.

14.5.6   Sewer Surcharge Rebate programme

ICI customers may be eligible for a rebate on a portion of the sewer surcharge. The sewer surcharge rebate applies to water not discharged into the sewer system, for example, water evaporated from cooling towers or used to make a product, and is credited on the water bill based on the sewer portion of the water rate paid.

14.5.7   Assistance for eligible low-income seniors and disabled persons

Since 2008, Toronto Water has a rebate programme for low-income seniors and low-income disabled persons. This rebate is set at the difference between the Block 1 and Block 2 rates, which represents a 30 percent reduction in their billing (based on the paid on or before due date rate). The rebate is only applicable if the household annual consumption is less than 400 cubic metres.

14.5.8   Partnering with retailers to sell water-efficient technologies and devices

In the past, Toronto Water worked with 'big box' commercial retailers such as Home Depot to ensure water-inefficient toilets were no longer sold in the Greater Toronto Area. To get the retailers on board, Toronto Water would hold a toilet sale at big box stores with utility employees on-site to help customers fill out their toilet rebate application. Appliances from the United States have water-efficient labelling already and Toronto Water has been advocating for the labels to be adopted in Canada. However, the difficulty is the federal government would be in charge of labelling, but it is the provincial government that determines the building code, and because of this multilevel oversight required, the bid has been unsuccessful to date.

14.5.9   Toronto's own water-labelling scheme

Toronto Water had a labelling scheme where customers would receive a $60 rebate if they purchased a product with a blue 'Save Toronto' label. However, the challenge is changing people's perceptions towards water-efficient products as they frequently believe they do not perform as well as standard products, for example, low-flow showerheads are perceived to have lower water pressure. To educate customers, Toronto Water set up a test protocol making sure each product with the 'Save Toronto' label was tested to increase credibility with customers. In 2011, this rebate programme was ended mainly due to big box stores no longer selling water-inefficient products and devices and the Ontario Building Code revision that increased water efficiency standards.

14.5.10   Distributing water-saving kits

Toronto Water gives away water-saving kits at city council-hosted 'Environment Days' where residents drop off inorganic waste for disposal. At these events the utility has a stall where it provides advice to customers on what water-saving devices can be installed to save water.

14.6   Communication and information demand management tools to achieve urban water security

14.6.1   School education and public awareness in the past

Toronto Water used to have an education programme where utility employees visited schools as part of the curriculum. In particular, the utility visited grade three classes and conducted presentations; however, this programme was cut due to budgetary constraints, as was a city environmental newsletter that contained a page on water-related issues. Regarding public awareness, Toronto Water used to have advertisement campaigns with visuals on poster boards and billboards and buses as well as radio ad campaigns. These campaigns were mainly related to toilet and washer rebate programmes or outdoor water efficiency. Today, Toronto

Water's main challenge is how to get people to understand water-related issues when the utility is competing with so many different city messages at any given time. To counter this, the utility is aiming to develop a one-stop website for customers to receive water-related messages, for example, what is stormwater and where does it go?

14.6.2 Education and awareness today

Toronto Water promotes water efficiency, rather than water conservation, as their customers do not wish to be labelled conservationist, as it is associated with being a hippie. As such, Toronto Water encourages its customers to use water wisely and efficiently. To better understand customers, the utility has participated in several studies that aim to profile individuals and determine which messages are best received and how.

Toronto Water's challenge in promoting water efficiency is the marketing and promotions are carried out in-house and tend to have a technical focus to them. However, for the message to be effective, it needs to be creative – using cartoons and common language – and focus its attention on the big picture. Toronto Water always makes a business case for customers to conserve water; however, the utility believes that sometimes the emotional case would last longer and have a greater effect, and so the utility is exploring the use of social marketing to test ways of emoting people and finding out what waste means to people, rather than relating water efficiency to saving money.

14.6.3 Promoting tap water: Water trailers

Toronto Water has created HTO To Go water trailers that serve as a fun and practical way of educating people about drinking water at public events. Between the months of May to September, the HTO To Go trailers attend select public events and is connected to the water supply so people can have a drink or fill up their water bottles while learning about the utility's programmes and services.

14.6.4 Billing inserts

Toronto Water uses water bill inserts to inform customers mainly on water rate changes. In the past, the utility had a water-watch leaflet that went out once a year and people tended to read that thoroughly.

14.6.5 Internet and social media

Toronto Water's website has a page devoted to helping residential customers conduct a home water audit and what to be aware of. The utility is also exploring the use of social media to raise water awareness. The challenge is only a limited number of people are engaged, and so the utility is trying to look at ways to get people engaged, even if it's just a 'murmur' in the background to let people know the utility exists.

14.6.6 Sharing lessons with other water utilities

Today the challenge is to raise awareness and increase customer engagement on a limited budget. The utility tries to cooperate with other municipalities on joint projects to share costs and be more efficient. Toronto Water also networks with other utilities through the Ontario branch of the American Water Works Association. The utility is on several committees that meet seven–eight times a year to keep abreast of progress in water management issues. As Toronto Water is often seen as a leader in programming, the other utilities frequently ask what they can expect in a couple of years' time. For instance, as the city of Toronto reduced Toronto Water's budgets for water conservation programmes, the other utilities will wish to learn best practices and lessons learnt on how they can ensure they retain their water conservation programmes or modify them to ensure they appear too successful to cut.

14.7 Case study SWOT analysis

14.7.1 Strengths

Toronto Water is aiming towards achieving universal water metering, with AMRs being installed ward by ward. This will ensure accurate meter reads and reduced leakage in the water distribution system. Universal metering will also promote the wise use of water by domestic customers with water bills being based on actual reads each quarter.

The utility provides a series of subsidies and rebates for ICI customers to increase water efficiency in their operations provided they make permanent changes in their operations to reduce water consumption. This not only ensures the wise use of water but also enhances their competitiveness by lowering operational costs. In addition, the utility offers a rebate on the sewage component if water is reused, for example, in cooling towers.

To lower UFW, Toronto Water replaces each year a significant length of water mains. The utility prioritises which sections will be replaced based on a formula that in addition to water main data aims to minimise disruptions to the community. To further lower UFW, the utility is establishing a water loss reduction and leak detection team.

To increase water efficiency in homes, Toronto Water distributes water-saving kits to the public at council-run Environment Days and provides information on its website on how to conduct a home water audit. To promote the wise use of water, the utility has water trailers that attends public events to encourage the wise use of water and foster dialogue between customers and utility employees. Regarding framing of water conservation messages, the utility is participating in several studies to better understand customers and determine which messages are most appropriate for specific customer segments.

14.7.2 Weaknesses

While Toronto Water has successfully decoupled water consumption from population growth, lower water consumption levels have led to the utility facing significant budgetary constraints. The result has been Toronto Water discontinuing its water efficiency labelling programme, as well as associated rebates, that encouraged residential customers to purchase water-efficient appliances in Toronto.

School education programmes, that involved classroom visits by utility staff, and public awareness programmes, including advertisement campaigns on billboards as well as water-watch billing inserts, have been discontinued due to limited finances. Furthermore, the utility does not offer subsidies or rebates to encourage water efficiency in homes.

The utility only distributes water-saving kits to customers once a year. This hampers the ability to ensure wise use of water as well as reduce energy and chemical costs in treating water and wastewater.

Toronto Water does not utilise effectively public awareness campaigns that only require minimal budgets, such as developing case studies on ICI customers saving significant amounts of water from permanent changes, or social media campaigns that promote the wise use of water through competitions and awards.

By facing significant budget shortfalls, Toronto Water is not able to fully realise efficiency gains from lowering energy and carbon emissions in providing water and wastewater services.

14.7.3 Opportunities

To ensure financial sustainability, as well as environmental sustainability, Toronto Water could work with the city of Toronto to develop a new water rate structure that separates out the current water rate into three separate rates: water, wastewater and stormwater. The water charge could be structured to ensure a fixed charge provides revenue stability for the utility ensuring infrastructure is maintained while promoting the wise use of water. A separate wastewater and stormwater charge could bring in additional revenue, while large ICI users could be offered a rebate on lowering the amount of stormwater that enters the combined system: lowering energy and other associated costs of treating wastewater. This would provide additional revenue for the utility to maintain its infrastructure in addition to increasing the resilience of the city to urban flooding events that are projected to increase with climate change.

To lower UFW, which in turn lowers budgetary pressures, the utility could improve its customer relations by providing, in addition to its proposed one-stop online website for water, a dedicated call centre hotline for the newly created leakage reduction team so customers can inform the utility of a leak immediately. This will improve repair time of urgent water system infrastructure. This service could be promoted to ICI customers as well as be part of the home water audit Toronto Water currently promotes on its website.

To promote the wise use of water, Toronto Water could link the installation of AMRs with the distribution of water-saving kits to build stronger customer relations as well as lower the costs of providing water and wastewater services. In addition, the utility could create an online water audit app that allows customers to conduct a home water audit which leads to them being able to order water-savings kits. The utility could either distribute them for free or charge a subsidised rate for the kits that could be attached to the next water bill. For low-income and other vulnerable customers, the utility could provide them for free.

To lower the costs of promoting the wise use of water, and engage customers more in social media, Toronto Water could work with other water utilities in the Ontario region to develop coordinated messages across the province at different times of the year. By pooling together resources the utilities could hire creative agencies to better target different customer segments through refined framing of messages. In addition, the utilities themselves could set up their own Ontario Water Day to promote the wise use of water across the province through the distribution of water-saving kits and information on how to lower water consumption as well educate the public on urban flooding measures required.

To further promote the wise use of water by domestic as well as ICI customers and increase the utility's engagement level on social media, Toronto Water could create competitions, awards and even case studies of customers using water wisely. For instance, Facebook competitions could be created for young people, while awards could be given out to members of the community who have proactively raised awareness on using water wisely, while case studies could be developed on ICI customers successfully saving water. The utility could even provide an award for the best performing ICI in different sectors with the benefit for the customer being increased exposure of their business or operation.

14.7.4   Threats

Toronto Water faces interrelated threats to its operations in the future from declining water consumption placing pressure on the utility's long-term financial stability. This has led to the existing 10-year financial plan relying primarily on successive water rate increases to fund continued infrastructure investment while conforming to a pay-as-you-go financing strategy. As a result of declining water consumption and unrealised revenues, projects amounting to $1 billion have been deferred in the 10-year Capital Plan. Meanwhile, the utility faces additional infrastructural costs in managing urban flooding issues that will likely increase in the future with climate change. Finally, the utility faces increased regulatory control and oversight that will impact both operating and capital budgets. For instance, new water quality standards will include new enforcement activities and potential penalties for noncompliance.

Table 14.5 Demand management tools to achieve urban water security

Diffusion mechanisms	Tools	Description
Manipulation of utility calculations	Pricing of drinking and wastewater	Single water rate for all customers
		Water rate includes water supply, stormwater management and sewer fees
		Industrial water rate for ICI customers
	Subsidies/rebates	Industrial water rate to encourage water conservation
		ICI Capacity Buy Back programme
		ICI rebates on portion of sewer surcharge
		Low-income seniors/disabled persons can receive subsidised water bill
Legal and physical coercion	Metering	Toronto's Water Supply Bylaw mandates the installation of water meters at all properties
		Utility working towards universal metering
		New AMRs will be installed for free
		Existing manual meters will be replaced with AMRs
	UFW	Active water main replacement programme
		Priorities for water main replacements determined by a model
		Water loss reduction and leak detection team
Socialisation	Water efficiency labelling	Partnership with commercial retailers to ensure inefficient toilets are no longer sold
		Toronto Water has been advocating for US water efficiency labelling to be adopted
		Rebate for purchases of products with blue 'Save Toronto' label
	Water-saving kits	Water-saving kits distributed at 'Environment Days'
Persuasion	Public education	Visuals on poster boards, billboards and buses and radio ad campaigns
		Participation in studies that aim to profile individuals and frame messages better
		HTO To Go water trailers educate people at public events
		Water bill inserts inform customers on water rate changes
		Website helps residential customers conduct home water audit
		Raising water awareness on social media

Table 14.6 Barriers to further urban water security

Barrier	Description
Economic	While Toronto Water has decoupled water consumption from population growth, lower water consumption levels have led to significant budgetary constraints
	Limited finances have led to the utility discontinuing several water conservation initiatives
	Toronto Water is not able to fully realise efficiency gains from lowering energy and carbon emissions in providing water and wastewater services
Infrastructural	The utility is having to raise water rates to ensure adequate investments are made in the water system
Technological	The utility only distributes water-saving kits to customers once a year hampering the promotion of using water wisely
Demographic	School education programmes involving classroom visits by utility staff have been discontinued
	Public awareness programmes, including advertisements and billing inserts, have been discontinued

14.8 Transitioning towards urban water security summary

Toronto Water uses a portfolio of demand management tools to achieve urban water security (Table 14.5). However, there are numerous barriers identified by the utility in achieving further urban water security in Toronto (Table 14.6).

Notes

1. TORONTO WATER. 2015. How is lake water turned into drinking water? [Online]. Available: http://www1.toronto.ca/city_of_toronto/toronto_water/files/pdf/water_treatment_process.pdf (accessed 12 May 2016).
2. CITY OF TORONTO. 2002. Water efficiency plan [Online]. Available: https://www1.toronto.ca/City%20Of%20Toronto/Toronto%20Water/Files/pdf/W/WEP_final.pdf (accessed 12 May 2016).
3. CITY OF TORONTO. 2015b. City of Toronto budget summary [Online]. Available: http://www1.toronto.ca/City%20Of%20Toronto/Strategic%20Communications/City%20Budget/2015/PDFs/GFOA%20Public%20Book%202015%20Toronto%20Budget%20compressed.pdf (accessed 12 May 2016).
4. CITY OF TORONTO. 2015a. 2015 water and wastewater rates and service fees [Online]. Available: http://www.toronto.ca/legdocs/mmis/2015/ex/bgrd/backgroundfile-77554.pdf (accessed 12 May 2016).
5. CITY OF TORONTO. 2015c. Service level review public works and infrastructure committee presentation [Online]. Available: http://www.toronto.ca/legdocs/mmis/2015/bu/bgrd/backgroundfile-85365.pdf (accessed 12 May 2016).

15 Vancouver transitioning towards urban water security

Introduction

Vancouver's transition towards urban water security focuses on ensuring the city has the best drinking water of any city in the world while reducing per capita water consumption. This case study analyses how Vancouver's water utility uses a portfolio of demand management tools to modify the attitudes and behaviour of water users to achieve urban water security.

The case study first provides a brief company background of Vancouver's water utility, along with an overview of the city's water supply and water consumption levels before discussing the city's strategic vision for achieving urban water security. The case study will then analyse the various demand management tools used by the city's water utility in an attempt to achieve urban water security before discussing the numerous barriers identified by the utility in achieving further urban water security in Vancouver.

15.1 Brief company background

Metro Vancouver is the regional government, and the city of Vancouver is a municipality within it. Metro Vancouver is a political body and corporate entity operating under provincial legislation as a 'regional district' and 'greater boards' that deliver regional services, planning and political leadership on behalf of

Urban Water Security, First Edition. Robert C. Brears.
© 2017 John Wiley & Sons, Ltd. Published 2017 by John Wiley & Sons, Ltd.

24 local authorities, which include the city of Vancouver. Metro Vancouver owns and operates the water supply, treatment and regional water supply system, while municipalities own and operate the local water distribution systems to supply water to residents and businesses. Metro Vancouver and member municipalities work together to supply clean, safe drinking water to more than 2.3 million people and businesses in the Metro Vancouver region.

As a member municipality of Metro Vancouver, the city of Vancouver's Waterworks Utility purchases bulk treated water and operates a citywide transmission and distribution system to deliver water to over 100 000 properties within Vancouver. The Waterworks Utility is self-funded. Revenues collected each year completely offset the costs to build and maintain the water system over the same period. Of the total budget of $108.6 million in 2014, $71 million was used to purchase bulk water from Metro Vancouver and the remaining $37.6 million was spent rebuilding and maintaining the water system.

The three core functions of the Waterworks Utility are ensuring drinking water delivered to customers meets all relevant health and quality guidelines, ensuring water system assets are well managed and resilient and making sure progress on the city's water consumption and water quality targets adopted as part of the Greenest City Action Plan's Clean Water goal effectively offsets population growth through efficient water use.[1]

15.2   Water supply and water consumption

Vancouver's drinking water comes from the Capilano, Seymour and Coquitlam reservoirs, which are protected and managed by Metro Vancouver. These watersheds collect surface water from rain and snowmelt and all three are closed to the public with no recreational, agricultural and/or industrial activities permitted within the watershed boundaries. Metro Vancouver is responsible for source water quality monitoring and treatment to ensure high-quality water is delivered to its member municipalities. Water treatment by disinfection destroys disease-causing or pathogenic organisms and secondary chlorine disinfection downstream of the watersheds helps prevent bacterial regrowth in the distribution system. Metro Vancouver is responsible for both primary and secondary treatments. The city of Vancouver does not further treat the water.[2] The Waterworks Utility is responsible for the operation and maintenance of its water distribution system that includes more than 1400 kilometres of water mains, 101 000 service connections and 16 000 metres. The Waterworks Utility also maintains a Dedicated Fire Protection System (DFPS) that consists of 11 kilometres of 600 millimetre diameter steel pipelines designed to withstand the maximum credible seismic event for Vancouver. Over half the consumers of water in Vancouver are domestic customers and a quarter listed as industrial, commercial and institutional (ICI) (Table 15.1).

Table 15.1 Vancouver water use by sector

Sector	Water use (percent)
1 and 2 family	30
Multifamily building	26
Industrial, commercial, institutional	25
System leakage	11
Parks	4
Other (including city-owned properties)	4

15.3   Strategic vision: Clean water and lower consumption

In 2005, the Board of Greater Vancouver Water District (GVWD) approved the Drinking Water Management Plan (DWMP) for Metro Vancouver and its member municipalities, which was updated in 2007 to incorporate the management of the source watersheds. The DWMP comprises three goals: goal one, provide clean, safe drinking water in which Metro Vancouver and its municipalities are committed to providing reliable access to adequate quantities of clean, safe drinking water to customers of Metro Vancouver; goal two, ensure the sustainable use of water resources to ensure the region can continue to grow and prosper while sustaining quality of life and the environment. As part of this goal, Metro Vancouver and its municipalities are committed to pursuing demand management strategies where using water more sustainably will contribute to economic prosperity, community well-being and environmental integrity. Goal three is to ensure the efficient supply of water as this optimises capacity and defers the need for new infrastructure and new water supply sources, in addition to renewing and replacing the region's ageing water transmission and distribution systems in an affordable way.

As part of DWMP municipalities will reassess the merits of developing residential water metering programmes and municipal rebate programmes for water-efficient fixtures and appliances; update municipal bylaws, utility design standards and neighbourhood design guidelines to enable and encourage on-site rainwater management as possible so it can be used for non-potable purposes such as irrigation; renew and replace ageing infrastructure to maintain required levels of service; undertake cost-effective leak identification and repair programmes; and implement where feasible and appropriate pressure reduction or pressure management programmes to reduce leakage and extend the life of infrastructure. Regarding efficiency, municipalities will enhance lawn-sprinkling regulations to address both season and peak day consumption issues.[3]

The Greenest City Action Plan is the city of Vancouver's strategy for being a leader in urban sustainability. The action plan involves multiple stakeholders, including the Council, residents, businesses, other organisations and all levels of government working together to implement the plan. The vision of the Action Plan is to build a strong local economy, vibrant and inclusive neighbourhoods and an internationally recognised city that meets the needs of generations to come. The Action Plan has 10 goals, with goal 8 being clean water. The goal is for Vancouver to have the best drinking water of any city in the world. Quality-wise Vancouver will meet or beat the strongest of British Columbian, Canadian and appropriate international drinking water quality standards and guidelines. Regarding quantity of water, Vancouver will reduce its per capita consumption by 33 percent from 2006 levels. Since 2006, there has been a 16 percent decrease in total water consumption in Vancouver.

The highest priority actions for 2011–2014 include water metering for new homes; developing and implementing enhanced water education; incentive and conservation programmes, including incentive programmes for low-flow toilets and increased education; and enforcement of lawn-sprinkling regulations. To date, the city has achieved a 16 percent reduction in water consumption from 2006 levels, from 583 litres per person per day to 490 litres per person per day: over halfway towards Vancouver's 2020 goal.[4]

15.4   Drivers of water security

The drivers of Vancouver's strategic vision for achieving urban water security include population growth, a lack of water storage and climate change.

15.4.1   Population growth

The Waterworks Utility is aiming to decouple population growth from water consumption – the city of Vancouver's population is projected to increase from just over 600 000 in 2006 to 740 000 in 2041 – to ensure the city can keep within its current storage capacity for drinking water.

15.4.2   Infrastructure: Lack of storage

One of the challenges of ensuring a sustainable supply of water for Vancouver is the city lacks adequate water storage capacity. Despite precipitation levels being high, the city is only able to capture around 20 percent of that rainfall which then becomes the city's drinking water. As such, if the city can prolong the lifespan of its three different watersheds, it can reduce potential capital costs of expanding the water supply network to meet increased demand: a challenge given water consumption approximately doubles during the summer period.

15.4.3   Climate change

In British Columbia, average annual temperatures have warmed by 0.5–1.7 degrees Celsius in different regions of the province during the twentieth century, with some parts of British Columbia warming at a rate of more than twice the global average. Over the past 50–100 years, British Columbia has lost up to 50 percent of its snowpack and total annual precipitation has increased by about 20 percent.[5]

The three watersheds Vancouver relies on for drinking water are expected to provide adequate water until 2050. However, climate change may impact rainfall and snowfall patterns that supply these watersheds: Vancouver's drinking water comes from mountain reservoirs and the snowpack melt and rainfall that supply them. April 1 snowpack has decreased on average by 25 percent in British Columbia over the past 50 years with some sites experiencing a 50 percent reduction.[6] Vancouver recognises that expanding the water supply or finding a new one is financially and ecologically expensive, and so water conservation is the best way to live within the city's means and avoid the need for source expansion. A summary of the projected climatic changes Vancouver will experience is listed in Table 15.2.

15.5   Regulatory and technological demand management tools to achieve urban water security

15.5.1   Price of water

During the rainy season, when the city's water supply is at its peak (November through May), all metered customers – residential and ICI – are charged an off-peak volumetric rate (Table 15.3). During the drier months, rates increase by around 25 percent to reflect the added cost of supplying scarce water to the city. Regarding wastewater, the Waterworks Utility charges a volumetric rate for sewage (Table 15.4). In addition, the utility charges a flat annual rate for unmetered domestic customers (Table 15.5).

15.5.2   Metering

All ICI customers and multifamily residential complexes are metered; however older single-family and dual-family homes are unmetered. Effective January 2012, all new single-family and two-family homes must have water meters (AMRs) installed and will move to volume-based pricing of their water usage. In 2014, the Waterworks Utility installed 1150 new water meters. On the residential side, multiunit residential buildings are metered by building, not by unit. This makes conservation efforts difficult as multifamily residential use accounts for around 26 percent of the city's total water use.[7]

Table 15.2 Projected changes in climate for Vancouver

Climate variable	Summary of change	Anticipated changes
Rainfall	Increase in average annual precipitation with a decrease in summer	*Averages* Increase of 6–9 percent in winter and a decrease of 14–15 percent in summer *Wet days* By the 2050s, precipitation during extremely wet days is expected to increase 28 percent relative to baseline of 1971–2000 *Extreme events* By the 2050s, a daily rainfall event that occurred once every 25 years will likely occur almost 2.5 times as frequently
Temperature	Increase in average annual temperature	*Averages* Annual increase of 1.7 degrees Celsius by the 2050s and 2.7 degrees Celsius by the 2080s *Warm days* Summer days above 24 degrees Celsius are projected to occur more than twice as frequently in the 2050s compared to baseline 1971–2000 *Extreme events* In the 2050s, an extreme heat event that occurred once every 25 years in the past will likely occur over three times as frequently
Sea level	Rising seas	*Averages* The province recommends using 0.5 metre global sea level increase to 2050, 1.0 metre to 2100 and 2.0 metres to 2200 *Extreme events* Sea level rise will cause damage when experienced together with storm surges
Extreme events	Increase	An increase in extreme events is projected including windstorms and heavy rainfalls

CITY OF VANCOUVER. 2012. *Climate change adaptation strategy*, Vancouver, BC [Online]. Available: http://vancouver.ca/files/cov/Vancouver-Climate-Change-Adaptation-Strategy-2012-11-07.pdf (accessed 12 May 2016)

Table 15.3 Metered seasonal rates

Period	2013 rate per unit	2014 rate per unit	2015 rate per unit
October 1 to May 31	$2.304	$2.385	$2.480
June 1 to September 30	$2.887	$2.988	$3.108

One unit equals: 2831.6 litres, 100 cubic feet

Table 15.4 Metered sewer rates

Period	2013	2014	2015
Year-round	$1.842	$1.906	$2.021

Table 15.5 Flat utility rates: water and sewer annual flat rates

Property type	2013	2014	2015
Single-family dwelling water (sewer)	$528.00 ($287.00)	$546.00 ($297.00)	$568.00 ($314.00)
Single-family dwelling with laneway house	$716.00 ($387.00)	$741.00 ($400.00)	$771.00 ($424.00)
Single-family dwelling with suite	$716.00 ($387.00)	$741.00 ($400.00)	$771.00 ($424.00)
Single-family dwelling with suite and laneway house	$904.00 ($487.00)	$936.00 ($504.00)	$973.00 ($535.00)
Strata duplex (per dwelling unit)	$358.00 ($194.00)	$371.00 ($201.00)	$385.00 ($213.00)

Waterworks Utility customers receive a water bill every 4 months. The amount includes a basic charge to cover costs of billing, meter maintenance and future meter replacement and a consumption charge based on the amount of water used during the 4-month period. Customers with AMRs have their water meter date read wirelessly from devices mounted to Waterworks Utility vehicles that drive through the streets every billing cycle.

The Waterworks Utility has not implemented a universal metering programme as the cost of implementation far outweighed the benefit due to water in Vancouver being relatively inexpensive. Nonetheless, the utility will re-evaluate this decision over time to see if it makes financial sense to have a universal metering programme. In addition, if customers want a water meter, the utility can install one for them; however it is extremely cost-prohibitive so very few customers voluntarily request a meter.

15.5.3 Reducing unaccounted-for water

In 2013, total water losses were estimated to be 12 billion litres, or 11 percent of total billed water purchased, a value of $7.2 million. To reduce system leakage and realise operational savings as well as be better stewards of water, the Waterworks Utility has a water loss reduction strategy that consists of the following listed programmes (Table 15.6).

Distribution Main Replacement programme

Since the early 1980s, Vancouver has conducted an annual water main replacement programme to manage the frequency and impacts of water main failures and improve the system's reliability. Since 2003, the target replacement rate for the distribution system has been 0.8 percent based on analysis regarding the lifespan of the city's water mains. In 2013, approximately 8.8 kilometres, or 0.6 percent, of

Table 15.6 Water loss reduction programmes

Programme	Description
Proactive leak detection survey	Distribution mains, services
Hydrant leak detection survey	Each hydrant in the city is checked twice annually
Reactive leak detection survey	Based on resident or corporate feedback
Pressure management	Pressures in certain parts of the city are lowered to reduce system leakage

CITY OF VANCOUVER. 2014. *Vancouver Water Utility annual report*, Vancouver, BC

distribution mains was replaced at a cost of around $8.6 million. The 2013 replacement rate of 0.6 percent was below the long-term target of 0.8 percent in response to a stable break rate over the past few years.

Upgraded sewers to prevent contamination

Vancouver replaces around 10 kilometres of combined sewer pipe each year with a separated storm sewer and sanitary sewer system. The old sewers are combined so in drier weather stormwater and waste are carried to the wastewater treatment plants together; however during heavy rains high volumes of stormwater can exceed the capacity of the system resulting in excess capacity overflowing directly into the city's waterways. The city aims to eliminate this sewage overflow by 2050 with a separated system that helps improve water quality, supports wildlife, increases biodiversity and reduces sewage backups during storms.

15.5.4   Alternative water sources

There is a policy on grey water – reusing water – but it is ill-defined in the city's building bylaw. There have been customers approaching the Waterworks Utility to install grey water or rainwater harvesting systems. To implement these systems they have to make an 'alternative solutions' application with the license and inspection group with each application assessed on a case-by-case basis. The incentive for having alternative systems installed is external to the city, for example, if a building is going for an LEED standard, or another green building certification, the building would get points for using rainwater or grey water.

15.5.5   Water restrictions on residential lawn sprinkling

Metro Vancouver's GVWD Water Shortage Response Plan (WSRP), which applies to all municipalities, restricts residential lawn sprinkling at all WSRP stages (1–4). In stage 1 June 1 to September 30, even-numbered addresses can water their lawns on Monday, Wednesday and Saturday mornings 4–9 a.m., and odd-numbered addresses on Tuesday, Thursday and Sunday mornings 4–9 a.m. In stage 2 even-numbered addresses can water their lawns on Monday morning only between 4 and 9 a.m., and odd-numbered addresses on Thursday morning only between 4 and 9 a.m. During stages 3 and 4, all forms of watering using treated drinking water are prohibited. These WSRP restrictions do not apply to the use of rainwater, grey water or any forms of recycled water.

15.5.6   Rebates for laundry machines

The Waterworks Utility is in partnership with local energy utilities to promote water (and energy)-efficient laundry machines. As part of this rebate, customers receive a rebate of $125 with the utility contributing $25 towards the total rebate amount. The rebate amount is received as a $125 reduction in the customer's energy bill and a $25 reduction in their water bill.

15.5.7   Subsidised indoor water-saving kits

The city of Vancouver offers indoor water-saving kits for customers. Each household can purchase up to two subsidised kits for $12 each (including taxes) with the retail value approximately $30 each. The water-saving kits are especially suited for older homes enabling customers to retrofit their kitchens and bathrooms, saving up to 20 percent less water and 15 percent on their water heating bill. The indoor water kits include the following:

- One 'Earth Massage' self-cleaning showerhead with adjustable spray setting from gentle needle to forceful jet
- One dual-setting Touch Flow kitchen aerator with swivel action for effective cleaning and choice of aerated jet and wide spray
- Two faucet sink aerators with solid brass casing and polished chrome finish, plus flow control constructed of long-lasting plastic
- Two toilet tank bags that are easy to install and made of noncorrosive materials that are resistant to microbes and fungal growth
- Two packages of leak detection dye tablets for testing leaks from the toilet tank or toilet bowl, which could save customers from wasting up to 150 litres of water per day
- One roll of Teflon tape to prevent leaks at hose connections

15.5.8   Installing water- and energy-efficient fixtures in restaurants

In 2015, the city partnered with FortisBC and BC Hydro to install new water- and energy-efficient dish-cleaning pre-rinse spray valves and faucet aerators in restaurants. These fixtures use up to 80 percent less water enabling restaurants to reduce their hot water use and energy bills. In total, 509 spray valves and 1591 aerators were installed in 476 participating restaurants. The estimated annual water savings are more than 100 000 cubic metres.[8]

15.5.9   Pilot toilet retrofit project

In the past, the utility conducted a pilot toilet retrofit project in which utility employees went through an old building being refurbished for apartments replacing the old toilets with new high-efficiency toilets, with the old toilets crushed and the porcelain reused. The utility was then able to monitor the building's reduction in toilet water consumption and work out the payback period of doing this retrofit, which was calculated to be around 4 years.

15.5.10   Water audits for ICI customers

The Waterworks Utility provides free audits for ICI customers, provided they sign a memorandum of understanding with the utility in which they agree to implement retrofits with payback periods of 2 years or less. The Waterworks Utility deliberately kept the payback period short to start a dialogue with ICI customers on what else can be done if basic retrofits, including the replacement of old spray valves, faucet aerators and toilets, can save significant amounts of water.

The Waterworks Utility will eventually provide case studies of the ICI customers that have taken part in the programme: currently the utility is in the middle of developing the case studies as the participating ICI customers have received their final reports and are now in the monitoring stage of determining the actual amount of water saved. The writing of case studies benefits both parties as it provides the business with free publicity, while the Waterworks Utility can show real-life examples of customers saving water.

15.6   Communication and information demand management tools to achieve urban water security

15.6.1   School programmes: H2 Whoa!

The Waterworks Utility has for the past 17 years contracted a theatre group to tour elementary schools and perform a play on water conservation. The plays involve 'slapstick' comedy that the children find very engaging, with high energy levels and fun. The plays are also purposely written for teachers to bring into the classrooms as the utility wanted to have a way of following up lessons learnt from the plays.

15.6.2   Public education: Promoting 'water-wise' gardening practices

The Waterworks Utility has found awareness campaigns are successful when they involve one-on-one contact with a customer or small group. To date the utility has successfully run a garden parties programme to promote water-wise gardening practices. The garden parties programme was a 3-year programme (2012–2014) that connected residents with a professional garden consultant who came to people's homes to deliver hands-on workshops. In total, 130 homes were visited and 650 residents received information on how to achieve a beautiful garden while using water efficiently. Similar to a Tupperware party, garden party hosts could invite friends and neighbours to the parties to share the experience, build a sense of community as well as learn about water-efficient landscape management in an engaging way.[9]

15.7 Case study SWOT analysis

15.7.1 Strengths

To encourage water conservation during the summer months, the Waterworks Utility charges a seasonal rate for all types of customers that is around 25 percent higher than the winter volumetric rate to reflect the added cost of supplying water to the city. The utility also charges domestic and nondomestic customers a volumetric sewage rate to encourage less consumption of water.

To reduce system leakage and realise operational savings, as well as be better stewards of water, the Waterworks Utility has implemented a proactive water loss reduction strategy that surveys distribution mains and hydrants in addition to receiving prompt customer feedback on leaks. The Waterworks Utility also replaces a portion of the water main system each year to reduce leakage.

Each year the Waterworks Utility replaces a portion of its combined sewer pipe with a separated storm sewer and sanitary sewer system. During heavy storm events this reduces excess capacity overflowing directly into the city's waterways. In addition to the separated system enhancing ecosystem health of the city's waterways, less wastewater will require treatment, reducing operational and maintenance costs of the Metro Vancouver's wastewater treatment plants.

To reduce peak demand during summer, the Waterworks Utility, in addition to restricting outdoor water use, has run garden parties to encourage neighbourhoods and communities to come together on water-wise gardening. In particular, Tupperware-style parties were held on how to achieve a beautiful garden while using water efficiently.

Indoor water conservation kits are available for customers to purchase at a subsidised price. These water-saving kits enable customers to retrofit their kitchens and bathrooms, saving up to 20 percent water and 15 percent on their water-related electricity bill.

The Waterworks Utility provides free audits for ICI customers provided they sign a memorandum of understanding with the Waterworks Utility in which they agree to implement retrofits with short payback periods. In the near future the Waterworks Utility will develop case studies on how ICI customers have made significant water savings from retrofits. This will provide the business with free publicity, while the utility can show customers real-life examples of businesses doing their part in saving water.

15.7.2 Weaknesses

The Waterworks Utility has not implemented a universal metering programme as the cost of implementation far outweighs the benefit due to water in Vancouver being relatively inexpensive. In addition, if customers do request a meter, the cost of purchase and installation is prohibitive resulting in a long payback period. Nonetheless, with the city lacking adequate water storage capacity, universal metering is critical in conserving water.

The Waterworks Utility has a limited school education and public awareness programme that focuses on school plays and garden parties to encourage the wise use of water. Specifically, the Waterworks Utility does not provide classroom education for a range of ages nor hold a variety of events targeting different customer segments throughout the year to promote the wise use of water.

Despite climate change projected to lower precipitation levels significantly, the Waterworks Utility is limited in its ability to promote alternative water sources including grey water and rainwater harvesting. When customers do wish to install an alternative system, they face a significant bureaucratic process with each application assessed on a case-by-case basis. As such, the Waterworks Utility does not offer financial incentives to develop alternative supplies.

15.7.3 Opportunities

The Waterworks Utility charges all customers a volumetric sewage rate. With the city operating a combined sewer system, the utility could provide financial incentives to ICI customers to implement sustainable urban design systems that capture and filter stormwater, reducing the volume that enters the sewage system resulting in a lower sewage-related bill.

The Waterworks Utility can expand its restaurant retrofit programme to other industries, such as hotels, and increase awareness of the free water audit programme available to ICI customers. This will increase the number of case studies available for the utility to publicise, further encouraging water conservation by all users. In addition, the Waterworks Utility can also link these case studies with energy consumption to enhance awareness of the link between water and energy bills. Finally, the utility could initiate industry awards to encourage businesses to make significant water savings.

The pilot toilet retrofit project could be expanded to all public institution buildings in Vancouver to enhance water savings. This would show the population that the city of Vancouver is proactive in saving water and achieving the 33 percent reduction target. To make further water savings, the Waterworks Utility could develop a subsidy programme for ICI customers to have toilet retrofits done.

In addition, the Waterworks Utility could partner with home equipment and appliance stores in Vancouver to offer customers an on-site rebate for water-efficient toilets.

The Waterworks Utility could expand its rebate programme to encourage residential customers to purchase water- and energy-efficient laundry machines and other home appliances. This could be done in partnership with local energy utilities, further enhancing awareness on the link between water and energy bills.

In Vancouver, the Waterworks Utility could initiate a grey water or rainwater harvesting project that educates the public on its environmental and economic benefits: the purpose being to encourage a change in the bylaw. If successful, the utility could then offer water efficiency funding to ICI customers installing alternative water systems, further enhancing the city's ability to reach its water conservation target.

The Waterworks Utility could distribute water-saving kits to unmetered single-family and dual-family homes to encourage water conservation. To reduce the financial cost of doing so, the utility could partner with local energy utilities as well as Metro Vancouver to share the costs. For metered residential customers, the utility could provide billing information on how much water can be saved from installing water-saving devices and offer customers the ability to order indoor water-saving kits online, over the phone or by mail and even offer to add the $12 charge to their next water bill. For low-income families in unmetered or metered households, the Waterworks Utility could subsidise the full costs of the kits.

15.7.4   Threats

Vancouver has limited water storage capacity and with population growth placing pressure on limited water resources and many customers unmetered, it will be a challenge in the future to ensure water supply meets demand. Specifically, climate change is projected to decrease precipitation levels and snowmelt during the summer months increasing the urgency of linking customers' water use directly with their water bills through water meters.

The city will need to increase investments in its water distribution system to lower its UFW rate as significant water, and financial, savings can be made from lower leakage levels. This will ensure every drop of water is being utilised: critical given demand could easily outstrip supply due to population growth and climate change reducing water availability.

With climate change projected to lead to wetter winter months, and Vancouver lacking adequate water storage capacity, the challenge will be to raise awareness on the need to conserve water year-round. This will require the Waterworks Utility to link water conservation with infrastructure savings because if customers do not conserve water during wetter months, the city will face significant capital costs of having to expand the storage capacity to ensure demand and supply balance year-round.

Table 15.7 Demand management tools to achieve urban water security

Diffusion mechanisms	Tools	Description
Manipulation of utility calculations	Pricing of drinking and wastewater	All metered customers charged a seasonal price for water
		All metered customers charged a sewer rate
		Unmetered customers charged a flat rate
	Subsidies/rebates	Subsidised indoor water conservation kits
		Rain barrel subsidy for residents
		Subsidy for water (and energy)-efficient laundry machines
Legal and physical coercion	Metering	All ICI and multifamily complexes fully metered
		From 2012 all new single-family and two-family homes must be metered
	UFW	Four-part water loss reduction strategy to lower UFW
		Targeted water main replacement programme
	Restrictions	Lawn-sprinkler use is restricted during summer months
	Alternative sources	City has a Dedicated Fire Protection System ensuring firefighters have access to water after earthquakes
	Separated systems	Replacement of combined sewer pipes each year to eventually create a separated system
Socialisation	Water-efficient technologies	Indoor water conservation kits
		Installation of water-efficient fixtures in restaurants
		Pilot toilet retrofit in old building
		Water audits for ICI customers
Persuasion	Public education	H2 Whoa! school plays
		Garden parties to encourage water-wise gardening
Lesson drawing and emulation	Case studies	Case studies of ICI customers and how they made water savings from retrofits

Table 15.8 Barriers to further urban water security

Barrier	Description
Economic	The Waterworks Utility has not implemented a universal metering programme as the cost of implementation far outweighs the benefit
Regulatory	When customers wish to install alternative systems, they face significant bureaucratic processes with each application assessed on a case-by-case basis
Demographic	The Waterworks Utility has a limited school education and public awareness programme that focuses on school plays and garden parties to encourage the wise use of water
	The Waterworks Utility does not provide classroom education for a range of ages nor hold a variety of public events throughout the year to promote conservation

15.8 Transitioning towards urban water security summary

Vancouver's Waterworks Utility uses a portfolio of demand management tools to achieve urban water security (Table 15.7). However, there are numerous barriers identified by the utility in achieving further urban water security in Vancouver (Table 15.8).

Notes

1. CITY OF VANCOUVER. 2014. Vancouver Water Utility annual report, Vancouver, BC.
2. Ibid.
3. METRO VANCOUVER. 2011. Metro Vancouver Drinking Water Management plan [Online]. Available: http://www.metrovancouver.org/services/water/WaterPublications/ DWMP-2011.pdf (accessed 12 May 2016).
4. CITY OF VANCOUVER. 2014. Vancouver Water Utility annual report, Vancouver, BC.
5. LIVESMART BC. 2015. Effects of climate change [Online]. Available: http://www. livesmartbc.ca/learn/effects.html (accessed 12 May 2016).
6. CITY OF VANCOUVER. 2012. Climate change adaptation strategy [Online]. Available: http://vancouver.ca/files/cov/Vancouver-Climate-Change-Adaptation-Strategy-2012-11-07.pdf (accessed 12 May 2016).
7. CITY OF VANCOUVER. 2015. Greenest city 2020 action plan: 2014–2015 implementation update [Online]. Available: http://vancouver.ca/files/cov/greenest-city-action-plan-implementation-update-2014-2015.pdf (accessed 12 May 2016).
8. CITY OF VANCOUVER. 2015. Greenest city 2020 action plan: 2014–2015 implementation update [Online]. Available: http://vancouver.ca/files/cov/greenest-city-action-plan-implementation-update-2014-2015.pdf (accessed 12 May 2016).
9. CITY OF VANCOUVER. 2015. Greenest city 2020 action plan: 2014–2015 implementation update [Online]. Available: http://vancouver.ca/files/cov/greenest-city-action-plan-implementation-update-2014-2015.pdf (accessed 12 May 2016).

16 Sharing the journey: Best practices and lessons learnt

Introduction

From the case studies of Amsterdam, Berlin, Copenhagen, Denver, Hamburg, London, Singapore, Toronto and Vancouver transitioning towards urban water security through the use of demand management tools, a series of best practices and lessons learnt have been identified for other cities around the world attempting to achieve urban water security. This chapter first outlines a series of best practices identified before discussing lessons learnt. Finally, the chapter discusses how water managers need to consider a range of demand management instruments classified under their respective diffusion mechanisms to achieve further urban water security.

16.1 Best practices

From the case studies a series of best practices have been identified for water utilities around the world implementing demand management strategies in an attempt to achieve urban water security.

16.1.1 Pricing water to promote conservation while ensuring revenue stability

Utilities with water tariffs that include both fixed and variable components have some form of revenue stability during times of reduced water consumption. This is essential to meet operating and maintenance costs of the water supply network.

Urban Water Security, First Edition. Robert C. Brears.
© 2017 John Wiley & Sons, Ltd. Published 2017 by John Wiley & Sons, Ltd.

One utility charges nondomestic customers with higher pollution content additional costs for the removal of suspended solids, nitrogen and phosphorus, while another is charging a volumetric sewage rate to encourage less water consumption. To promote water conservation during summer months, a couple of utilities charge both domestic and nondomestic customers a seasonal rate to reflect the added costs of supplying water over the summer period. Meanwhile, some utilities use taxes to promote water conservation, ensuring the costs of providing water and wastewater services are recovered.

16.1.2   Universal metering key to water conservation

Utilities with universal metering of all domestic and nondomestic customers, including submetering of individual apartments in complexes, can directly communicate with all water users on the need to use water wisely. In addition, metering of all customers means the ability to deploy a computerised billing system that automatically detects abnormally high or low consumption in the water distribution system. Utilities that do not have universal metering are moving towards having all customers metered with AMRs installed, ensuring accurate meter reads and reduced leakage in the water distribution system. Water utilities are also exploring the use of smart meters to reduce water consumption and enable accurate leak detection.

16.1.3   Investments in the water distribution system key to lowering UFW

To reduce UFW, utilities have proactively invested in the maintenance and upgrading of the water distribution system. To detect leaks, utilities commonly survey the whole system over a designated period of time, ranging from once a year to one that is staggered over a 5-year period. In addition, utilities with low UFW rates have proactively upgraded or replaced their water mains and pipes. To increase the efficient use of water resources, one utility has a computerised main replacement system that identifies existing and potential problem areas for early replacement. Another utility prioritises which sections of water mains will be replaced based on a formula that also seeks to minimise disruptions to the community. To build better customer relations, utilities have established 24/7 service centres for customers to report leaks to.

16.1.4   Reducing energy and carbon emissions

Utilities are promoting the wise use of water to achieve energy savings from treating less wastewater, which in turn lowers carbon emissions. By treating less wastewater it also means wastewater treatment plants have additional capacity during heavy storm events, reducing overflows into waterways. The same utilities are implementing renewable energy solutions – including wind turbines and the recovery of heat and energy at wastewater treatment plants – to further reduce energy costs and reduce carbon emissions. Decentralised systems are also providing opportunities to reduce energy and carbon emissions from treating and distributing potable water with vacuum toilet and sewage systems along with greywater. A couple of utilities have invested in the development of a separated storm sewer and sanitary sewer system to reduce combined system overflows during heavy storm events. In addition to a separated system enhancing ecosystem health of waterways, sewage water that enters wastewater treatment plants will be more concentrated, reducing operational and maintenance costs of treating waste.

16.1.5 Source protection: Reducing treatment costs

Water utilities are concerned about the quality of their source water. Controlling pollutants at their source, in contrast to removing them in the drinking water treatment process, not only reduces human health risks but also reduces treatment costs. One utility has found that nearly all its groundwater wells located inside the city's boundaries are contaminated from agricultural and industrial pollution, and so it relies almost exclusively on groundwater supplies from outside the city's boundaries. To protect these supplies the utility has entered into partnerships with local farmers in the planting of forests and reduction in pesticide use to protect groundwater supplies. Another utility, which relies on sourcing water from inside the city's administrative boundaries, enforces stringent rules on which types of activities are permitted inside zones around each well. Other utilities source their drinking water from outside their city's boundaries in river basins that are prone to environmental degradation. One utility in a river basin prone to forest fires is partnering with a government agency to work on maintaining forests to reduce fires, and associated soil erosion, ensuring adequate water quality.

16.1.6 Targeted subsidies

Utilities have either developed targeted incentive programmes for specific user segments or developed blanket incentive programmes for all customers, both domestic and nondomestic. One utility has identified public housing as being one of the largest consumers of water and so has developed a targeted toilet subsidy programme to replace old toilets with more efficient ones. Another utility has a range of subsidies and rebates for nondomestic customers who agree to make permanent changes in their operations to reduce water consumption: in addition to saving water, it also enhances their competitiveness by lowering operational costs. In comparison another utility offers both residential and commercial customers a range of rebates on devices and technologies as well as a free water audit to promote water efficiency.

16.1.7   Promoting water efficiency

To encourage water efficiency, utilities commonly encourage the installation of water-efficient devices in homes. To ensure installation is done correctly, one utility has partnered with plumbers, other utilities and nonprofits to visit homes and install the devices professionally. Another utility has established an awareness programme that provides water-saving kits and a water audit checklist for students to conduct audits at home. One utility distributes water-saving kits to the public at council-run environmental days in addition to providing information online on how to conduct a home water audit. Another utility sells subsidised indoor water conservation kits enabling customers to retrofit their kitchens and bathrooms, saving both water and energy.

Promoting water-efficient appliances based on a labelling scheme helps consumers make informed sustainable choices. One utility has created a dual voluntary and mandatory labelling scheme for water-using devices and appliances, while another utility provides incentives for both domestic and nondomestic customers to purchase water efficiency labelled devices and appliances. Another way of promoting water efficiency, which also reduces energy costs and carbon emissions, is water utilities initiating in-house corporate social responsibility initiatives. With water utilities installing water-efficient devices and promoting in-house conservation, it means they can understand the challenges first-hand of modifying attitudes and behaviour towards water, as well as other resources.

16.1.8   Water conservation becoming a way of life

Water conservation requires behavioural change. Recognising this, one utility promotes water conservation throughout the year, not only during times of drought, to ensure water conservation becomes a way of life and culturally ingrained in people. Another utility, recognising the disconnection between people and nature, informs customers on how their water is sourced and the environmental impacts of overconsumption: specifically, the water awareness campaign informs customers living near surface water on where their drinking water is coming from and how overconsumption can impact the recreational use of the rivers. To encourage community participation in saving water, one utility has established voluntary groups to work with utility employees on educating households in neighbourhoods on water conservation. Meanwhile, other utilities participate in public events by hosting stalls or setting up novel displays that encourage a dialogue between the public and utility employees on water conservation.

16.1.9   Demographic-targeted messaging

To better target customers, one utility has identified demographic groups that are most receptive to accepting the installation of water efficiency devices. Using this data, the utility is deepening its engagement with these groups by implementing demographic-specific campaigns. At the same time the utility recognises there are various demographic groups that it is not reaching, and so in the future the utility will widen its demographic-specific messaging campaigns. Another utility is participating in research studies to better understand customers and determine which messages are most appropriate for their customer segments. Utilities also target specific demographic groups to promote water conservation as a way of life. For instance, one particular utility, targeting gardeners, has successfully organised garden parties to encourage neighbourhoods and communities to come together on water-wise gardening practices.

16.1.10   Nondomestic water-saving plans

Nondomestic users are often the largest customer segment for water utilities. One utility has introduced water efficiency management plans for nondomestic customers to understand their water usage and identify areas to reduce consumption and increase efficiency. This scheme is mandatory for all large-scale customers with annual submissions of plans as well as the installation of private meters at all major water-using areas. Other utilities offer nondomestic customers free audits. As part of this service, one utility has nondomestic customers sign a memorandum of understanding in which they agree to implement retrofits with short payback periods.

16.1.11   Recognising water savings

Providing recognition for outstanding contributions towards water conservation is an important tool for raising public awareness. One utility offers an award for protecting and raising awareness of water resources with recipients considered role models who inspire others to take ownership of the water and ensure the long-term sustainability of water resources. Another utility is developing case studies of nondomestic customers that have made significant water savings from retrofits, providing the business with free publicity while showing other nondomestic customers real-life examples of how businesses have saved water and lowered their operational costs.

16.2   Lessons learnt

From the case studies a series of lessons learnt have been identified for water utilities around the world implementing demand management strategies in an attempt to achieve urban water security.

16.2.1   Pricing water too cheaply

The price of water needs to recover the full economic cost of providing water and water-related services as well as promote conservation. However, there is strong public pressure to ensure prices are kept low to protect customers. One utility faces a price ceiling on how much it can charge consumers for water, and so it must rely on non-price demand management tools to promote water conservation to achieve its water consumption target. Meanwhile, there is pressure to ensure water pricing does not impact the competiveness of businesses, with one utility having a water tax for heavy water users that actually decreases with consumption, despite the city attempting to attract sustainability-related companies. This hinders the utility's ability to encourage the wise use of water by domestic users, as nondomestic customers are 'rewarded' for higher consumption. One utility faces the problem that when the utility discusses retrofits with its customers after conducting a water audit, customers rarely request retrofits if the payoff period is greater than 18 months as the price of water is considered 'cheap'. The structure of water prices can also have an adverse effect on a utility's ability to operate. If water consumption is too low, utilities will have less revenue to cover the fixed costs of operating and maintaining the water distribution system. One utility has an inclining block rate for residential customers that has remained unchanged for more than two decades; however, because of increased water efficiency and water conservation over time, the majority of the customers never leave the first block, impacting conservation efforts and the utility's revenue. In addition to falling consumption levels reducing revenue for maintaining and upgrading the water distribution system, including reducing UFW, utilities can face difficulties servicing their long-term debt.

16.2.2   Lack of universal metering

Universal metering ensures customers pay for the exact amount of water consumed, promoting water conservation and reducing leakage in the system. However, when customers are not fully metered, it hinders the utility's ability to promote water as well as energy conservation. Several utilities have universal metering of all customers both domestic and nondomestic; however city regulations do not require the submetering of individual apartments in buildings. One utility is faced with the challenge that despite every 'customer' having a water meter, the actual population that is metered is very low due to the majority of the city's inhabitants living in rented apartments that do not have submeters, inhibiting the utility's ability to communicate directly to all water users on the need to use water wisely. Another utility, despite having universal metering, has no submetering of multifamily apartment buildings, impacting the utility's ability to directly link customer's water consumption with water bills. This will likely become an issue for this particular utility in the future as the number of people living in multifamily apartments in the city is projected to increase. A couple of utilities are rolling out progressive metering programmes to ensure all customers within a specified timeframe will be metered, while another has a water meter subsidy programme to achieve submetering of all apartments. However, without universal metering that includes submetering of apartments, utilities will continue to face a challenge in promoting water conservation, detecting leaks and recovering the full costs of providing water services.

16.2.3   Inability to develop alternative sources

Alternative water supplies, such as greywater and rainwater harvesting, reduce the need to increase supply to meet rising demand. However, many of the utilities face regulatory hurdles in developing these systems. One utility faces a national ban on the use of greywater systems, despite these systems reducing water consumption levels and energy costs in providing non-potable water. Nonetheless, the same utility's customers can use rainwater for toilet flushing; however, this requires special permission from the health inspectorate to ensure there are no cross-connections: a lengthy process resulting in minimal numbers of rainwater harvesting systems being installed. Another utility faces significant regulatory hurdles on promoting greywater and rainwater harvesting systems despite the city setting a goal on sourcing water from secondary water resources. In addition, this particular utility's non-domestic customers are barred from developing private wells for use in industrial/commercial operations because the city will only approve wells if the water is of drinking water quality. Meanwhile, another utility, despite climate change projected to lower precipitation levels, has difficulty in encouraging the development of alternative sources due to the city only approving greywater and rainwater harvesting systems on a case-by-case basis. Utilities also face a lack of alignment of policies across levels of government with one utility located in a state that allows the use of greywater but only if the local government has approved its use, which it has yet to do, limiting the utility's ability to conserve water.

16.2.4   Not fully utilising subsidies

Incentive programmes encourage the uptake of water-efficient devices and appliances; however, many utilities are not using the full range of subsidies and rebates available to encourage water efficiency and conservation. For instance, one utility that encourages the wise use of water does not use financial incentives to promote water-efficient technologies such as low-flow showerheads or fix leaking toilets. Another utility, despite projecting an increase in domestic consumption over the next two decades, does not provide subsidies or rebates for the purchase of water-efficient toilets or household appliances. Meanwhile, another utility, faced with increasing nondomestic demand, does not offer rebates to either nondomestic, or domestic customers on the purchase of water-efficient devices and appliances for domestic uses of water. Finally, one utility, despite offering commercial rebates on water-efficient devices, does not offer commercial customers subsidies for the installation of water-efficient technologies in industrial operations that save verified amounts of water.

16.2.5   Limited education and public awareness

Education is required to change the attitudes and behaviour of people towards water; however, utilities often have limited educational and public awareness programmes. For instance, one utility's school education and public awareness programme revolves around a yearly school play and the hosting of garden parties to encourage wise water use by gardeners. This is despite the utility facing drier summers from climate change in addition to a rising population. Another utility fails to maintain a wide range of educational programmes for both schools and the public to increase the emotional message of needing to save water. Specifically, its education programme only targets young children rather than deepening education on water-related topics as children progress through the educational system. In addition, the same utility does not hold year-round education workshops on specific water-related topics or partner with nonprofits to spread targeted messages. Finally, utilities are often failing to target specific demographic segments. One utility recognises it is only 'speaking' to a very small part of their overall customer base, yet the utility is failing to target young people who are believed to be one of the largest users of water, but have no insight into the price of water as they are not the bill payer.

16.2.6   Lack of funding

It is common for utilities to have their water conservation budgets linked to revenue generated from the sale of water. However, a decrease in water consumption impacts the financial viability of future water conservation and efficiency programmes. One utility has decoupled water consumption from population growth only to face significant budgetary constraints from significant revenue losses. The result has been the utility discontinuing its water efficiency labelling programme as well as associated rebates that encouraged residential customers to purchase water-efficient appliances in the city. In addition, the utility's school education programmes, that involved classroom visits by utility staff, and public awareness programmes, including advertisement campaigns on billboards as well as billing inserts, were discontinued due to limited finances. Finally, the same utility faced a budgetary shortfall in the maintenance of the water distribution system.

16.2.7   Lack of online presence

Utilities commonly provide water-saving kits for distribution to customers; however, the scale is limited the majority of the time to outdoor environmental events and purchase at local government offices or through small-scale distribution by volunteers. Utilities are not utilising the Internet and social media to encourage customers to purchase these kits or provide an ability for them to order online via web portals.

16.2.8   Unsuitable infrastructure

One utility recognises that its current infrastructure is very inflexible in meeting new challenges, in particular managing higher volumes of stormwater in the combined sewage system. If the utility did decide to build a new water pipeline network, the utility will face a peak in depreciation costs. Furthermore, other assets supporting the operation and maintenance of these pipes will need to be depreciated resulting in large financial losses for the utility. Another utility faces challenges to its infrastructure from rising groundwater levels due to reduced water consumption and increased water efficiency resulting in potential flooding of the city's infrastructure.

16.2.9   Lack of political will

One utility believes that for a broad range of climate adaptation measures to be implemented, it requires coordination and political will as the implementation will occur in existing processes of planning and decision-making which involve many stakeholders including planners, investors, politicians and even the law body: all of which are difficult to coordinate without significant political will.

16.3   Moving forwards

Water managers need to consider a range of demand management instruments classified under their respective diffusion mechanisms to achieve further urban water security.

16.3.1   Manipulation of utility calculations

Utilities need to ensure they recover the full economic cost of providing water, including the operational and maintenance costs of the water supply system. With water conservation reducing revenue, utilities need to ensure their pricing structure maintains some form of revenue stability. In addition, utilities need to ensure there is enough capital to invest in upgrading infrastructure. This can be achieved through the use of fixed and variable charges with the fixed charge ensuring there is adequate revenue for future capital projects, while the variable charge provides customers an incentive to conserve water. Utilities with variable seasons can explore the use of seasonal charges to reflect the scarcity of supply during warmer months. In addition, utilities that also maintain wastewater treatment plants can structure their tariffs to provide incentives for customers to reduce the amount of waste entering the sewage system, which in turn lowers the operational and maintenance costs of wastewater treatment plants.

In order to encourage water conservation, the wise use of water or reductions in the cost of providing water and wastewater services, utilities should increase their use of subsidies and rebates in promoting water efficiency in homes and businesses. For instance, utilities can provide subsidies for customers to buy water-efficient devices for their homes such as tap faucets and efficient showerheads. Utilities can use subsidies to encourage the purchase of low-flow toilets where customers receive a coupon to be redeemed at stores. Alternatively customers can receive a rebate after having purchased water-saving devices that is deducted from their water bill. Utilities can use subsidies and rebates to promote water-efficient appliances such as washing machines as this will reduce water and energy consumption levels and lower carbon emissions. In addition, utilities need to use subsidies or rebates to encourage the installation of alternative supply systems, particularly with heavy water users such as businesses, schools and universities, as this will reduce pressure on limited supplies as well as reduce energy usage and carbon emissions in providing potable water. Finally, utilities should encourage customers to reduce the amount of rainwater entering the wastewater treatment system by, for example, having stormwater retention ponds installed on private property to capture rainwater before recharging groundwater supplies. This will save energy costs from treating less wastewater and reduce infrastructure costs of having to build larger wastewater treatment plants to deal with higher volumes of wastewater from more frequent climate change-related heavy rainfall events.

16.3.2   Legal and physical coercion

Utilities need to significantly increase their use of ordinances in reducing water consumption levels, for example, working with the local government, to make it mandatory that all new developments and renovated properties have water-efficient devices such as low-flow toilets, efficient showerheads and taps with maximum flow levels installed. This will lead to reduced energy consumption levels and lower levels of carbon emissions, as a significant amount of the energy bill is associated with the heating of water. Utilities should increase their use of alternative sources of water such as rainwater harvesting or greywater for flushing toilets or watering gardens. This will lead to utilities being able to match demand for water with the appropriate quality of water because currently the majority of utility customers use water of potable standards to flush toilets and water gardens. The benefit of developing alternative sources of water is that it lowers the treatment costs of providing drinking water and wastewater (and in most cases reduces carbon emissions). Utilities should ensure universal metering of all customers, including submetering of apartments, as water consumption levels decrease significantly when customers are metered as they can see directly how their behaviour towards water impacts their water (and energy) bills.

16.3.3   Socialisation

Utilities can develop, with their respective local government, water labelling schemes for water-efficient devices and household appliances as a way to conserve water/use it wisely or reduce energy costs (carbon emissions) from providing water and wastewater services. Utilities should show leadership in water conservation/using water wisely by enacting in-house water conservation strategies. The benefit of doing so is utilities will be able to identify more easily barriers to water conservation strategies based on what works/does not work in-house when promoting water conservation with employees. The utilities can develop partnerships with other utilities – energy and gas companies – to provide an overall message of the need to conserve resources as water and energy bills are interlinked. The partnerships can extend into joint messages or providing water-saving devices or conservation materials to their respective customers. The utilities should also develop services such as household visits to private homes and apartment buildings to help occupants find ways of reducing water such as fixing leaky pipes, installing water-saving devices and increasing people's awareness of the need to conserve water/use it wisely. Finally, the utilities can utilise GIS software and automated billing software to detect inefficient water use from which customers can be offered a water audit service to reduce their water consumption levels.

16.3.4 Persuasion

The utilities should improve their use of classroom materials to encourage young people to conserve water as they are usually heavy users of water and establish lifetime habits at a very young age. The utilities can increase their use of open days to encourage more school children and members of the public to visit water and wastewater treatment plants as this increases awareness of how potable water is produced, the costs of providing it and the costs of treating waste. Utilities should increase their use of billing inserts as this is usually the only direct contact the utilities have with their customers. These inserts can inform customers of sudden increases in water consumption levels that could be due to leaks and provide information on how to contact the utility for water-saving devices and water conservation tips. Finally, utilities need to utilise demographic data to refine the framing of their water conservation messages to achieve their water-saving targets. In particular, utilities in the future could have a range of messages that target different ages, ethnic or socioeconomic groups with messaging.

16.3.5 Competition/emulation/mimicry

The utilities should encourage the ability of water users to compare their water use with other users in the same neighbourhood, suburb, city or state, for example. Though household composition is different in each household, the ability to compare their use with the average provides a strong incentive to conserve as it is 'what others do'. With increased data from AMR and smart meters, water utilities could in the future compare households of similar composition (number of occupants, private house/apartment, etc.) to enable comparisons with similar uses. In addition, the utilities can develop competitions between communities to encourage water conservation, such as between schools, with the winners receiving prizes. Water utilities can also foster competition between users such as holding competitions for young people who save the most water in their households by changing household habits towards water. The utilities should develop case studies of individuals, households or businesses that have successfully saved water. This provides other water users with best practices on how they can save water. The utilities can use role models such as young people who are winners of water prizes, schools that have saved the most water or community leaders to encourage the wise use of water. Finally, the utilities should develop partnerships with nonprofits as they have the ability to reach a variety of customer segments. For instance, nonprofits can be used to distribute water conservation materials or water-saving devices while spreading their own message.

Conclusions

In traditional urban water resources management, water managers forecast population growth and economic development to determine future levels of demand. If there is a projected supply deficit (demand outstripping supply), water managers rely on large-scale water supply projects to transport water over large distances to bridge the deficit. However, these supply-side solutions have become unfavourable due to their environmental, economic and political costs. Environmentally, supply-side solutions impact the availability of water for ecosystems. Economically, the reliance on distant water increases not only the costs of transportation but treatment costs too. Politically, scarcity of water is likely to lead to inter-user, inter-sectoral, inter-regional and international competition, or even conflict, over scarce water supplies as water resources typically cross internal and external political boundaries. In addition, traditional urban water resources management fails to account for uncertainty in supply from climate change extreme weather events.

In transitions towards urban water security, water managers aim to balance rising demand for water with limited supplies. This is achieved through the use of demand management strategies that promote water conservation during times of both normal and atypical conditions and through changes in practices, culture and people's attitudes towards water resources. Demand management strategies are both antecedent and consequential where antecedent strategies attempt to influence the determinants of target behaviour before the performance of the behaviour, while consequential strategies influence the determinants of target behaviour *after* the performance of the behaviour.

To achieve urban water security, water utilities will need to transition towards the sustainable use of water that balances demand with supply. In this book, a transition is defined as a well-planned shift from one sociotechnical system to another where a sociotechnical system is a stable configuration of human and non-human elements including technology, regulations, market and user practices,

Urban Water Security, First Edition. Robert C. Brears.
© 2017 John Wiley & Sons, Ltd. Published 2017 by John Wiley & Sons, Ltd.

cultural meanings, infrastructure, maintenance and supply networks. Specifically, a transition is a change in the way society operates and occurs through a combination of behavioural, cultural, ecological, economic, institutional and technological developments that positively reinforce one another for change to occur. In transitions, institutions create a futuristic vision of the new sociotechnical system and then coordinate the appropriate resources (economic, financial, knowledge, etc.) to achieve that vision. Transitions occur over multilevels: at the macro level (landscape), meso level (regime) and micro level (individuals). The macro level is the exogenous environment the system operates in and is beyond the direct influence of the meso and micro levels. It is relatively static and includes the institution's goals and visions that guide transitions at the meso level. The meso level comprises the sociotechnical system's regime: if a transition is to be successful, institutions must change, in a coordinated way, the norms and values of the regime's social users. At the micro level, innovations are tested against one another. If these innovations are successful, they will branch out and attract mainstream audiences.

Before a transition can occur, there first needs to be a misfit or 'gap' between individuals' and society's deeply held values and the current conditions they face. At the macro level, institutions can create tension with the meso level (regime) by creating a gap between the new strategic vision of the future and the current regime's outdated practices. At the micro level, institutions can place pressure on the meso level through innovations that attempt to create a gap between a new alternative regime and the current outdated regime. Transitions can also be triggered by changes in the external environment, in particular social, technological, economic, environmental and political changes. For a transition to occur – the closing of these gaps, there needs to be force applied. Supportive forces are top-down (macro-level) forces that standardise practices or routines through standards and directives. Formative forces are bottom-up (micro-level) forces that create pressure on the regime through innovations that challenge the existing regime. Formative forces can be artificially created by institutions.

In transitions, the application of supportive forces at the macro level can take the form of alternative visions of the future, while at the micro level formative forces can be in the form of diffusion, which is a process where ideas, norms and innovations are communicated over time among members of a social system. Diffusion in the context of sustainability involves the adoption of new environmental innovations that initiate social change in the structure and functions of society towards the environment and its natural resources.

In transitions towards the managing of water sustainably to achieve urban water security, a transition is a well-planned, coordinated transformative shift from one water system to another over a long period of time, where a water system is comprised of physical and technological infrastructure, cultural/political meanings and societal users. In a water system, society is both a component of the water system and a significant agent of change in the system, both physically and biologically. The main drivers in transitions towards new sociotechnical systems in water are rapid population growth and urbanisation, rapid economic growth and rising income levels and increased demand for energy and food as well as climate change which leads to scarcity of good quality water of sufficient quantity for all water users and uses.

Transitions towards urban water security involve a transition from first- to third-order scarcity. In traditional water resources management (first-order scarcity management), managers mitigate the impacts of both variations to, and rising demand for, water resources by increasing supply. These supply-side solutions have typically consisted of large-scale dams, reservoirs and pipelines transporting water over large distances. Over time these traditional supply-side solutions have become unfavourable due to their environmental, economic and political costs. In second-order scarcity management, the focus is on increasing economic and technological efficiency in the management of water resources. In particular, attention is placed on the economic value of water, resulting in the pricing of water to manage demand. However, second-order scarcity policies eventually have to give way to third-order policies because they do not address the main driver of water scarcity: human behaviour. In third-order scarcity, water managers combine second-order scarcity policies of economic and technical efficiency with demand management policies that focus on changing people's norms and values towards the environment in general and water in particular.

In transitions towards third-order scarcity, there are two components to modifying the attitudes and behaviour of water users (domestic and non-domestic) at the meso level: first, there is the strategic or macro-level sustainability vision or goal – the water-saving target – and second, there is the operationalisation of this strategy at the micro level. In this transition, the application of supportive forces at the macro level can be in the form of targeted levels of water consumption (e.g. per capita litres per day) with the baseline for comparison being current levels of (unsustainable) water consumption, while at the micro level, using the definition of diffusion, the application of formative forces is demand management, with two main types of instruments available for water managers to use to achieve urban water security: regulatory and technological instruments and communication and information instruments. Regulatory and technological instruments are frequently used in the management of water and involve setting allocation and water-use limits. In addition, regulatory and technological instruments are used to provide incentives for all water users to conserve water and use it efficiently. Meanwhile, communication and information instruments encourage a water-orientated society. In particular, communication and information tools aim to change behaviour through public awareness campaigns around the need to conserve scarce water resources.

This book examined transitions towards urban water security in nine leading cities around the world. Specifically, the book conducted case studies on how water utilities in Amsterdam, Berlin, Copenhagen, Denver, Hamburg, London, Singapore, Toronto and Vancouver use demand management tools to achieve urban water security. From the case studies a series of best practices and lessons learnt are developed for other utilities aiming to achieve urban water security through demand management. In addition, the book discussed how water managers need to consider a range of demand management instruments classified under their respective diffusion mechanisms to achieve further urban water security.

Best practices

To ensure water conservation efforts do not threaten revenue stability, several utilities have water tariffs that include both fixed and variable components. To ensure full cost recovery in treating wastewater, one utility charges non-domestic customers with high pollution content additional charges for treatment. Meanwhile, another utility charges customers a volumetric sewage rate to encourage water conservation and reduce treatment costs. Utilities are also using seasonal rates to promote water conservation during summer months.

Utilities that have universal metering, including submetering of individual apartments, can directly link customer's water usage with their water bills. This encourages water conservation throughout the year. Utilities are moving towards having all customers metered with AMRs, ensuring accurate meter readings and reduced leakage in the water distribution system.

To reduce UFW, utilities commonly survey the whole system over a designated period of time and proactively invest in their distribution system. To lower UFW, utilities have also invested in computerised systems that identify existing and potential problem areas for early replacement. In addition, utilities have also established service centres that allow customers to inform the utility of leaks anytime of the day.

Utilities are promoting the wise use of water to achieve energy savings from treating less wastewater, which in turn lowers carbon emissions. By treating less wastewater, it also means wastewater treatment plants have additional capacity during heavy storm events, reducing overflows into waterways. The same utilities are implementing renewable energy solutions to lower their energy costs and carbon emissions. Utilities are also developing alternative water sources, including greywater and rainwater harvesting, to reduce pressure on limited supplies, which in turn reduces energy and carbon emissions from treating and distributing potable water.

Water utilities are concerned about the quality of their source water as the higher the water quality, the lower the treatment costs. Utilities that rely on water within their own city's boundaries often face contamination challenges. To ensure source protection within city limits, one utility enforces rules on what types of activities are permitted inside zones around each well. At the river basin level, one utility is partnering with a government agency to reduce the risk of forest fires and associated soil erosion from impacting source water quality, while other utilities are working with farmers to reduce pesticide use that impacts groundwater quality.

Several utilities have developed incentive programmes for domestic and/or non-domestic customers. For instance, utilities offer domestic customers subsidies for water-efficient devices and toilets as well as water meters, while other utilities provide a range of subsidies and rebates for non-domestic customers who agree to reduce water consumption. Utilities are also offering all customers free water audits to increase water efficiency.

To encourage water efficiency in homes, one utility is partnering with other utilities to visit homes and install water-saving devices for free. Another utility provides students with water-saving kits and a water audit checklist so they can conduct water audits at home. Utilities also distribute water-saving kits at public events, for instance city council environmental days. In comparison another utility sells subsidised water-saving kits for homes. Water utilities are also practising what they preach by initiating in-house water conservation efforts. This means they can better understand the challenges of achieving water savings from the customers' perspective.

Utilities are promoting water-efficient labelled devices and appliances, helping consumers make informed choices on which products to purchase. To increase water efficiency, one utility has created a dual voluntary and mandatory labelling scheme, while another utility provides incentives for all types of customers to purchase water-efficient labelled devices and appliances.

To encourage behavioural change towards water, one utility promotes water conservation throughout the year, not only during times of drought, to ensure water conservation becomes a way of life. Another utility, recognising how people are becoming disconnected from the environment they live in, informs customers on where their water is sourced and the environmental/recreational impacts of overconsumption. To encourage community-wide behavioural change, one utility has established voluntary groups to educate households on water conservation. Meanwhile, other utilities host or participate in public events to encourage dialogue on water conservation.

To better target customers, one utility has identified demographic groups that are most receptive to accepting the installation of water-efficient devices and deepened its engagement with these customers. Meanwhile, another utility is participating in research studies to better understand customers and determine which messages are most appropriate for their customer segments. As non-domestic users are often the largest users of water, one utility has introduced mandatory water efficiency management plans for all large-scale customers. Other utilities offer non-domestic customers free audits. As part of this service, one utility has non-domestic customers sign a memorandum of understanding in which they agree to implement retrofits with short payback periods.

Providing recognition for outstanding contributions towards water conservation is an important tool for raising public awareness. One utility has annual awards for outstanding individuals or organisations that protect and raise awareness of water resources. This encourages others to take ownership of water and ensure its long-term sustainability. Another utility is developing case studies of non-domestic customers who have saved significant amounts of water through retrofits, providing the business with free publicity while showing other customers real-life examples of water savings.

Lessons learnt

The price of water needs to recover the full economic cost of providing water and water-related services as well as promote conservation; however, there is strong public pressure to ensure prices are kept low. One utility faces a price ceiling on how much it can charge consumers for water, and so it relies on non-price demand management tools to promote water conservation. Meanwhile, there is pressure to ensure water pricing does not impact the competiveness of businesses, with one utility having a water tax for heavy water users that decreases as consumption increases. When water is considered 'cheap', it can hinder conservation efforts: one utility, after conducting water audits for its non-domestic customers, has difficulties in encouraging them to request retrofits as the payback period is often too long. The structure of water prices can also impact a utility's ability to operate: if water consumption is too low, utilities face financial pressure to operate and maintain the water system.

When customers are not fully metered, it hinders the utility's ability to promote the wise use of water. Several utilities have universal metering of all customers, both domestic and non-domestic; however city regulations do not require the sub-metering of individual apartments. One utility is faced with the challenge that despite every 'customer' having a water meter, the actual population that is metered is very low due to the majority of the city's inhabitants living in rented apartments that do not require submeters. Without universal metering, including submetering of apartments, utilities will continue to face a challenge in promoting water conservation, detecting leaks and recovering the full costs of providing water services.

Many utilities face regulatory hurdles in developing alternative water supplies including greywater and rainwater harvesting systems. One utility faces a national ban on the use of greywater systems while rainwater harvesting for toilet flushing is allowed; however the installation involves a lengthy regulatory process. Another utility faces significant regulatory hurdles on promoting greywater and rainwater harvesting systems despite the city setting a goal on sourcing water from secondary water resources. Meanwhile, another utility has difficulty in encouraging the development of alternative sources due to the city only approving greywater and rainwater harvesting systems on a case-by-case basis. Utilities also face regulatory barriers across multiple levels of government with one utility located in a state that allows for the use of greywater only if the local government has approved its use, which it has yet to do so.

Many utilities are not using the full range of subsidies and rebates available to encourage water efficiency and conservation. For instance, utilities that encourage the wise use of water often fail to provide subsidies and rebates for water-efficient devices or household appliances. Utilities are also failing to provide incentives for non-domestic customers to conserve water in their operations.

Utilities often have limited educational and public awareness programmes to encourage water conservation. For instance, one utility's school education programme revolves around a yearly school play. Another utility fails to maintain a wide range of educational programmes for both schools and the public to increase the emotional message of needing to save water. In addition, the same utility does not hold year-round education workshops on specific water-related topics or partner with non-profits to spread targeted messages. Finally, utilities are failing to target specific demographic segments. One utility's research indicates that it is only reaching a very small part of their overall customer base. Specifically, the utility is failing to target young people who are believed to be one of the largest users of water, yet have no insight on the price of water because they do not pay the bill.

One utility has decoupled water consumption from population growth only to face significant budgetary constraints. The result has been the utility discontinuing its water efficiency labelling programme, as well as associated rebates, that encouraged residential customers to purchase water-efficient appliances in the city. In addition, the utility's school education programmes, which involved classroom visits by utility staff, and public awareness programmes, including advertisement campaigns on billboards as well as billing inserts, have been discontinued due to limited finances.

Utilities commonly provide water-saving kits for distribution to customers; however, the scale is limited for the majority of the time to outdoor environmental events, purchase at local government offices or through small-scale distribution by volunteers. Utilities are not utilising the Internet and social media to encourage customers to purchase these kits or provide a facility for them to order online via web portals.

Unsuitable infrastructure is becoming a challenge for water utilities. For instance, one utility recognises that its current infrastructure will struggle to manage higher volumes of stormwater in its combined sewage system. However, if the utility decided to build a new system, it would face a peak in depreciation costs. Meanwhile, another utility faces a costly challenge to its infrastructure from rising groundwater levels due to reduced water consumption and increased water efficiency.

Lack of political will hinders further efforts to conserve water and related resources. For instance, one utility believes that for a broad range of climate adaptation measures to be implemented, it will involve the coordination of many stakeholders, which is difficult to achieve without significant political will.

Moving forwards

With water conservation reducing revenue, utilities need to ensure their pricing structure maintains some form of revenue stability. In addition, utilities need to ensure there is enough capital to invest in upgrading infrastructure. This can be achieved through the use of fixed and variable pricing structures. Utilities with variable seasons can explore the use of seasonal charges to reflect the scarcity of supply during warmer months. In addition, utilities that maintain wastewater treatment plants can implement volumetric charges to reduce the amount of waste.

Utilities will need to significantly improve their use of ordinances in reducing water consumption levels. This can be achieved through partnerships with local government agencies to develop water efficiency standards for all new or renovated developments. The benefit is reduced water and energy usage and lower carbon emissions. The utilities need to work with government agencies to develop alternative sources of water, which in turn lower water and wastewater treatment costs. Utilities also need to ensure universal metering of all customers, including submetering of apartments, as water consumption levels decrease significantly when customers are metered.

Utilities need to increase their use of subsidies and rebates in promoting water efficiency in homes and businesses. In addition, utilities need to use subsidies or rebates to encourage the installation of alternative supply systems. Finally, utilities with combined sewage systems need to encourage customers to reduce the amount of rainwater entering the system. This will save energy costs from treating less wastewater and reduce infrastructure costs of building larger wastewater treatment plants.

The utilities can develop with their respective government agencies water labelling schemes for water-efficient devices and household appliances as a way to conserve water/use it wisely or reduce energy costs (carbon emissions). The utilities should also show leadership in water conservation by showing they are responsible users of water and associated resources and enacting conservation strategies themselves.

The utilities can develop partnerships with other utilities to provide an overall message of the need to conserve resources as water and energy bills are interlinked. The partnerships can extend into joint messages or providing water-saving devices or conservation materials to their respective customers. The utilities should also develop services such as household visits to customers to reduce water consumption, for example, detect leaks and install water-saving devices. Finally, the utilities can utilise GIS software and automated billing software to detect inefficient water use.

The utilities should improve their use of classroom materials to encourage young people to conserve water. The utilities can increase their use of open days to encourage more schoolchildren and members of the public to visit water and wastewater treatment plants to understand the water cycle. Utilities should increase their use of billing inserts to inform customers of sudden increases in water consumption levels. Finally, utilities need to utilise demographic data to refine the framing of their water conservation messages to achieve their water-saving targets.

The utilities should encourage the ability of water users to compare their water use with other users. With AMRs and smart meters, water utilities could in future enable households of similar composition to compare their water consumption. In addition, the utilities can develop competitions between communities to encourage water conservation. Water utilities can also hold competitions for young people to save as much water as they can. The utilities should develop case studies of all types of customers that have successfully saved water. This provides other water users with tips on how they can save water. The utilities can use role models such as young people to encourage the wise use of water. Finally, the utilities should develop partnerships with not-for-profits to reach different customer segments.

In conclusion, achieving water security is not a static goal; instead it is an ever-changing continuum that alters with numerous challenges. Therefore, future water security depends not only on meeting increased demand but also on how effectively humans can use limited water resources to meet these needs.

Index

Page numbers in *italics* refer to illustrations; those in **bold** refer to tables

Urban Water Security, First Edition. Robert C. Brears.
© 2017 John Wiley & Sons, Ltd. Published 2017 by John Wiley & Sons, Ltd.